GROW.
COOK.
EAT.
SHARE.

Copyright © 2019 by Caran Jantzen

ALL RIGHTS RESERVED. No part of this book may be reproduced or transmitted in any form by any means, electronic or mechanical, including photocopying and recording, or by any information storage and retrieval system, except as may be expressly permitted in writing from the publisher. Requests for permission should be addressed to The Homestead Press LLC, Attn: Rights and Permissions, P.O. Box 172, Elbert, CO 80106.

ISBN 978-0-9600259-1-6

Table of Contents

Dedication i
Glossary of Farm Terms iii
Introduction vii

GROW. Spring

The Arrival of Spring 3
Heritage Chickens 11
Rhubarb 23
Pasture Raised Pigs 31
Wild Spring Tonic 43
My First Kidding 49
The Potager Garden 57
Feed the Bees 67

COOK. Summer

High Summer 79
Farm Fresh Eggs 89
A Visit from the Vet 99
Grape Jelly 105
How To Do It All 113
The Farm Cats 121
Summer Cordial 129
The Straw that Broke the Goat's Back 137

EAT. Autumn

Autumn Days 147
Eat the Weeds 157
Poultry Menagerie 167
Sauerkraut 179
An Unlikely Farm Dog 187
Fresh Baked Bread 195
Bucks will be Bucks 203

SHARE. Winter

Winter Solace 215
Lard 225
Homemade Herbal Salve 235
Chicken Dinner 241
Winter Elixir 249
Food is Love 255
Ode to the Animals That Left too Soon 263
Spring Again 273
Acknowledgements 281

Dedication

This book is dedicated to my grandmothers, both of whom have passed from this life to their heavenly ones. To my mom's mom, Grandma Anne, you told me stories of your challenging life as a farmer's wife on the Canadian prairies. You truly lived the homesteading life, from milking cows, to tending your garden; from sewing your family's clothing, to cooking everything from scratch. You helped me bake, harvest in the garden, and crochet. I am so grateful for the homesteading spirit you inspired in me. To my dad's mom, Grandma Lena, you were a woman of incredible tenacity. You endured much pain in your life, and yet you pressed on with your family and in your faith. You continued to learn new skills, and were determined to be an independent woman to the end. I am grateful for the faith and strength you modeled to me. I miss you both, and I am so thankful I could call you both 'Grandma'. Until we meet again.

A Glossary of Less Familiar Animal and Homesteading Terms

Box - a structure to house bees; a box can be used for bees to produce and store honey, and can be called a honey box; or it can be used to raise offspring, and can be called a brood box.

Buck - a mature male goat.

Buckling - a young male goat; a male goat kid.

Cell - a six sided hole or pocket that makes up a comb; queen bees lay eggs in cells; worker bees also store honey in cells.

Cheap meat - meat that is produced in such a way that the true cost of feed, labor, etc., as well as the environmental cost, isn't reflected in the final retail price; Concentrated Animal Feeding Operation (CAFOs), feedlots, and factory farming here and abroad produce cheap meat.

Chicken tractor - a small portable chicken coop, that either has two or four wheels attached to it to move it easily. The purpose of portability is so the birds have access to fresh grass and bugs daily or every few days.

City clothes - the clothes a person wears when he or she does something away from the farm/homestead—hopefully the smell of animals is not lingering in these clothes!

City friend - a friend who lives in a city or town and who doesn't have the same appreciation for or understanding of a farm, homestead, or livestock that you do—and has likely never before experienced a goat nibbling on her cashmere sweater!

Cockerel - a cocky, male teenage chicken.

Colony - a group of bees that live together.

Comb - wax produced by honey bees and built in a hexagonal pattern of cells; honeycomb is the comb where the bees store their honey.

Cull, to cull - to end the life of an animal who is sick or injured; some breeders also cull animals who do not conform to breed standards.

Doe - a mature female goat.

Doeling - a young female goat; a female goat kid.

Drake - a mature male duck.

Ethically grown produce - produce that is grown in a way that considers the health of the soil it is grown in, as well as the health of the insects and animals that depend on it, and not just the vegetable itself.

Ethically raised meat - meat from animals that have been raised in their ideal conditions, and fed an optimal diet for their species, so that they will have the best life possible, as well as a painless, humane end.

Guinea fowl - a type of poultry related to the pheasant, that is recognized by its featherless head and neck, and its dark grey and white speckled feathers. It is most often raised for meat.

Hen - the term for mature female poultry, including chickens, turkeys, ducks, peafowl and guinea fowl.

Heritage animals - a designation for animals that have historically had a prominent place in agriculture, and whose pure breeding lines have been preserved. They have retained diverse, self-sustaining characteristics such as longevity, strong maternal instincts, natural breeding, etc.

Hoop House - an unheated structure consisting of a hooped or bowed frame made from metal or plastic, and covered in thick, clear plastic, to extend the growing season of vegetables and fruits by 4-6 weeks; it may be high enough to walk in or only high enough to cover the crop.

Keet - a guinea fowl chick.

Kid - a young male or female goat.

Menagerie - a mixed group of different types of animals or, in our case, a mixed group of poultry; traditionally a menagerie was to exhibit rare and unique animals.

Molt, Hard molt - a stage when chickens lose their old feathers and regrow new feathers; a hard molt is when a chicken loses most of its feathers at once before it regrows new feathers — and it looks like a drowned rat!

Muscovy - a breed of duck that is often raised for meat; this breed requires less water than most ducks and can be raised more easily on a small homestead because it can survive without a pond and only a basin of water.

Nose-to-tail eating - using as much of the animal as possible so as not to waste anything—we're getting close, although I haven't pickled pig's feet yet!

Paddock - a section of field fenced off for the animals.

Pasture-raised - animals raised with access to outdoor pasture daily, where they are free to eat grass, and bugs, and soak up the sun.

Peachick - a male or female peafowl chick.

Peafowl - consisting of male peacocks and female peahens.

Peahen - a mature female peafowl.

Potager garden - an expression describing a garden that is grown 'for the soup pot.'

Poult - a turkey chick.

Pullet - a young, female chicken that has just started to lay eggs.

Queen - a mature female cat.

Rut - the season that male goats are most able and likely to breed with female goats — and the season they stink the most!

Tom - a mature male turkey; a mature male cat; my amazing content editor and publisher!

Wether - a castrated male goat.

Introduction

Some people say that farming is in their blood. I don't quite understand how this could be, but I understand the sentiment. I've felt this feeling deep inside of me since I was a teen. My mom's parents were farmers in their early years, and suburban homesteaders once I knew them. My desire to homestead and farm the land evolved slowly over the course of many years, from a desire tucked in the recesses of my mind, to taking a more prominent place in my thoughts as the gruesomeness of factory raised and produced foods became more exposed, and I had my own hungry mouths to feed.

Two books inspired me early on in this journey toward raising and producing food for myself and my family: The Omnivore's Dilemma, by Michael Pollan; and Animal, Vegetable, Miracle, by Barbara Kingsolver. While the former is written as an ethical eater's manifesto, the latter reflects on the journey of one family in response to the ethical dilemma that mono-cropping and CAFO farming (if you can call it that) produces. Whatever excuses I had had up until that point evaporated. It was my turn to launch a peaceful rebellion by deliberately choosing ethical meat raised in a way that cares for the animals, and vegetables, grains and legumes produced in ways that care for the living soil in which it is grown.

Buying ethically raised meat and vegetables from small scale farmers whose hands we had shaken was our first step. My hubby and I wanted to know how the pigs, cows, and chickens that we would eat were raised from beginning to end. We wanted them to experience, as Joel Salatin puts it, the pigness of the pigs. But we still had such a consumerist mindset. We only wanted certain cuts of meat—we didn't yet know what nose to tail eating even meant; we wanted to eat certain vegetables year round. And we winced at the pricetag of these humanely raised foods, without realizing the true costs that this type of farming requires.

Now that we are raising our own animals for meat, I feel as if we eat it more consciously; we savor each cut; we use as much of the animal as possible. We know the value of the meat, because we knew the animal it came from, and how many trips down to the paddocks we made over the course of their lives. We haven't yet slaughtered and butchered our own animals, although we have watched our chickens be processed. Doing the processing ourselves is the next step.

Producing vegetables and fruits is much the same. We value the food we produce because we know how many times we have had to water and weed each crop. We know the lengths we went to to purchase organic, open pollinated seed, to amend the soil, and to naturally prevent pests. We have agonized over crop rotation charts, pruning techniques, companion planting information, and seed harvesting until it has begun to feel like we are treading familiar ground. The fruit and vegetables we produce have been watered in part by our sweat and our tears. I have come to begin to understand what Wendell Berry meant when he wrote: "The care of the Earth is our most ancient and most worthy, and after all our most pleasing responsibility."

It would be very eye opening for people to go through the cycle of raising to consuming, even just once. From sowing a seed to harvesting a crop; from watching a chick hatch, to thoughtfully and gratefully harvesting it. If at all possible, I believe that all people should have the experience of raising their own vegetables and animals. The point isn't to teach them to be farmers, but to teach them to be mindful consumers, grateful for the sun and the rain, and the farmers that work so hard, day in and day out, to feed their families, neighbors, and communities; and thankful for the happy lives the animals were given. For as Wendell Berry continues, "To cherish what is left of [the earth] and to foster its renewal is our only hope."

GROW.
Spring

The Arrival of Spring

"In spring, at the end of the day, your hands should smell like dirt."

- Margaret Atwood

Springtime in the garden is an arousal for the senses. Fragrant fruit tree blossoms delight the eyes and nose. The subtle buzz of worker bees beginning to gather pollen tickles the ears; swollen buds, preparing to burst open at the sun's prompting, bring hope to the eyes and the heart. The warmth of the sun-kissed dirt on the hands offers comfort. In springtime, there is nowhere I would rather be than in the garden.

On and around our homestead, the first to bloom is forsythia, willow, purple dead nettles, dandelions, and daffodils. Next come the maples, tulips, elderflowers, and peach tree blossoms. The rest of the fruit trees follow suit—plums and cherries, pears and apples; and with them bloom the rhododendrons, peonies, and periwinkle. Our overwintered broccoli and kale is also flowering—delicate yellow blossoms; I love adding them to salads, or eating them fresh off the plant. In fact, one of my egg customers has requested buying the flowering kale stalks for several years now.

Flowers and trees are not all that awaken with the warmer weather. Whorls of burdock and dandelion leaves pop up among the garden beds and on the lawn. The crinkled leaves of the rhubarb plants make their first appearance; and slender stinging nettle stalks begin to sprout in the cool, shady corners of the property. Leaves of mint, sage, catnip, and lemon balm begin to gather around the base of last year's twiggy growth. As I walk by, I often pause to stoop down and rub a leaf between my fingers, inhaling the pungent aroma it releases when it is crushed.

While beauty can be found in the darkest days of winter, nothing is more inspiring to me than seeing the green tips of the garlic beginning to push their way up through the soil after waking from a deep winter's sleep. There is nothing more peaceful than dreaming of the days when I can once again feel the warmth of the sun on my back as my hands work the soil. Longing for the taste of tangy, spring rhubarb and sweet, juicy strawberries is almost as delightful as actually eating them.

As spring officially begins, I find myself counting down the days to our last average frost date. I can hardly wait to begin another season of new life and growth. I feel excited as I anticipate planting seeds in the dirt and setting out the seedlings I started indoors. It never gets old. It's always miraculous when those first tender shoots push through the surface of the earth, stretching towards the sun.

So far, I have sown snap peas, hardy lettuces, cilantro, parsley, carrots, and beets in the ground. The peas and lettuces are just beginning to peek through the soil; I am still waiting for the rest of vegetables and herbs to spring to life. Indoors, I have a row of tender shoots growing in makeshift seed starter trays—onions, chives,

The daffodils emerge along the edge of the wooded area of our homestead each spring.

lettuce, broccoli, and cauliflower. For now, they are being pampered along my sunny southern facing window sill. Once they are bigger, they will need to be set outside during the day for several weeks to help them adjust to outdoor variables before planting, such as wind, rain, and fluctuating temperatures.

On sunny days, the chickens venture up into the orchard in search of tender young weeds and juicy bugs. A mat of green plantain covers the ground, and tender blackberry shoots spring to life in unexpected, new locations. As the earth dries out, the chickens like to stir up the dust, coating their feathers in an instinctual act called dust bathing. They can sit like this for hours, soaking up the warmth of the sun. Mr. Green Thumb, the sprouts, and I can end up with sunburns on these kinds of days if we aren't careful.

Springtime is also when precious little animals appear on the farm. Our adorable piglets were delivered to the farm last month and have not stopped eating or growing since their arrival. We have three adorable little doelings bouncing around the field next to their mama, and more are on the way. I've made a sheltered nook for the ducks, and they have begun to lay in a corner—I'm hoping that one of them will feel broody enough to sit on the clutch of eggs and hatch a few; I'd love to see little ducklings toddling along behind their mama.

As wondrous as spring is in the garden, it is not nearly as romantic as I have built it up to be in my mind throughout the silent winter months. It is also full of the chill of rain that can fall for a month on end, until there is finally a break in the clouds. Spring is filled with the heavy mud that clings to the soles of my rubber boots; once dry, it breaks apart and crumbles onto the floor of our back room when I pull my boots back on. Spring is soggy hay, and food troughs filled with rainwater. It is watching the orchard bloom, only to have nearly half of the blossoms washed away by an unexpected rain storm.

On walks around our property, and as we drive through patches of undisturbed nature, the kids and I look for wild, edible flowers and plants and take note of where they are so that we can return to harvest them on a sunny day. "Elderberry!" the girls shout from the back of the truck as we drive to their dance class. My herbal apothecary has no elderflowers left in it, as they have all been used up in cordials and teas; I am especially anxious to refill the empty jar I have sitting on the shelf. Dandelion blossoms are another seasonal flower I harvest each spring—their flowers make

Soil blocks are planted with a variety of spring vegetable seeds. A couple weeks after these seeds sprout, they will be ready to be planted in the spring garden.

a lovely infused oil, perfect for the salve, body butter, and soap I make throughout the year.

The sweet potato slips I have been tending have outgrown their little tray, and are ready for their own one gallon pots where they can stretch out their roots and their leaves. What started from three handsomely plump sweet potatoes has multiplied into over thirty sweet potato slips over the course of one month. Fifteen have gone into pots, which I carry outside every sunny morning for an hour or two. Another fifteen slips still need to be planted, and are patiently waiting on my sunny kitchen counter. Our summers don't stay hot enough for long enough to directly plant these slips in June; instead anyone who grows sweet potatoes in southwestern Canada needs to start these plants indoors, several months before we are able to plant them outside.

This past weekend, my son spotted a honeybee swarm through our front window. There were thousands upon thousands of bees, hovering in cloud formation, around one of the sour cherry trees in our front orchard. I frantically called my friend and beekeeping mentor who keeps several hives on our property. No response. I left a voicemail, and then sent a text, hoping she would get my message in time. I was certain that one of the honey bee colonies had split —I just wasn't sure how long they would hang around our property, or even the neighborhood.

Between chopping vegetables, and sautéing the meat for our supper meal, I continued to run over to the front window to check if the swarm was staying put. Over the course of about an hour, the bees settled down, and started to cling to the trunk of the tree, only about two feet off the ground. I breathed a sigh of relief and called my friend again, leaving another message stating that for now, the bees were settling in comfortably on one of our trees.

She and her husband came out after dark that evening—a little ping on my phone let me know they had arrived. My little sprouts, dressed in nightgowns and plaid pajamas, rushed downstairs to the front window and watched as the beekeepers' flashlights bobbed up and down among the trees. We could barely make out their outlines as they worked around the base of the tree. And no sooner had they come, then they were walking back up the hill, the light of their flashlights growing fainter, until it disappeared.

The next morning, my friend let me know that her hive had indeed split—the queen had left with roughly half of her colony, leaving queen larvae in her place. However, the newly established hive hadn't cooperated well enough the night before to move the

Mr. Green Thumb tends to his newly planted tomato and pepper plants in our unheated greenhouse.

box back up to where the rest of the hives were situated. She was planning to come again that night to move the box. To her relief, the bees had all settled in by nightfall, and she was able to move her box full of bees back to the top of the hill.

Just when the rain seems as if it will never end, and the mud puddle in the pig's pasture has reached record depth, the sun returns to warm the our corner of the earth. Slowly, the puddles recede, and more days than not are dry as I go about my chores. The plants in the garden beds have really started to grow now, and the seedlings will be ready to go in the ground next week. Spring is maturing, much like the bulbous green fruits that are lining the tree branches. Soon enough, summer will be upon us.

Heritage Chickens

"We can see a thousand miracles around us every day. What is more supernatural than an egg yolk turning into a chicken?"

- S. Parkes Cadman

For us, the first step toward becoming a homesteading farm was to own chickens. They're small enough to handle, and we did not feel all that intimidated by them. After poring over blogs, websites, and chat rooms, we were ready to take the leap. I figured that a chicken's overall intelligence level was about the same as my homesteading knowledge (no offense to our chickens); we would make a great team. After a bit of research, we decided that purchasing chickens from a hatchery was not for us. These conventional hens have been bred to produce as many eggs as physically possible, at the expense of many other chicken-like qualities. We instead chose old-fashioned, heritage breeds, known for their climate hardiness and for their ability to forage for food and hatch their own young. The chickens we bought were lovingly hand raised by several reputable breeders and 4-H members in the Fraser Valley.

In the end, we chose our first chickens as much for their showy feathers as for their chicken traits and for the eggs they would lay. Two were white with lacey black collars around their necks; a couple of them were all black with a green-blue sheen in the sunlight; a few had varying shades of gold and copper; two had intricate black and white patterns; and one had silky, lavender-grey feathers. They were an instant hit with our children, who claimed and named each bird. Our boys chose masculine names for their hens: Shadow, Midnight, Lightning, and Thunder; while the girls reused names they had already given their stuffed animals – Puffy, Fluffy, Dotty, Queenie, Lacey, and Goldie. And then there was Baldy, who the kids thought resembled a bald eagle.

Those several months after purchasing our eleven hens felt almost magical. Upon waking, the kids would rush barefoot outside into the sunshine, arguing over who would be first to feed, water, and cuddle the chickens. They took turns gathering, cleaning, and putting away the eggs, without giving a thought to the fact that they were actually doing chores. We quickly nicknamed our youngest the chicken whisperer. At two years old, she would enter the coop, swiftly scoop up a chicken under each arm, and rocking back and forth, she softly serenaded them into a drowsy stupor.

Every couple of days, my husband and I heaved the heavy, portable chicken coop across the small plot of grass in our front yard. The birds scratched and pecked at the fresh greens and bugs under their feet, along with their feed and kitchen scraps. As the weather grew warmer, the patches of lawn where the chicken tractor had been, stopped growing back. By the end of the summer, our entire lawn looked like a brown and yellow patchwork quilt, trimmed with a green border.

CJ loves cuddling all the animals on our homestead, including the chickens.

Not long after our chickens arrived, one of our hens, Puffy, went broody. In a conventional chicken farm, chickens are culled if they go broody, because this means they stop laying and focus instead on trying to hatch their clutch of eggs. But on our fledgling homestead, this was exciting news. However, we didn't have a rooster yet, and we didn't have the heart to tell little Puffy that her eggs would never hatch. Instead, we asked a neighbor, who also had a backyard flock, if we could purchase some fertilized eggs. Under the cover of darkness, my husband and I swapped out Puffy's warm, unfertilized eggs for the five we hoped would hatch—one cream, one brown, and three lovely, pale green eggs. Three weeks after the switch, Puffy became the proud mama to three little chicks, which we named Peep, Cheep, and Sleepy. Although hatching chickens under a hen is a sustainable way to farm, it isn't very efficient. So we decided to invest in a small, two dozen egg incubator to speed the process along. By that time, we had also purchased a rooster. We could not wait to hatch our own eggs from our mixed breed flock and see what unique colors and patterns would emerge. Peep, Cheep, and Sleepy were already laying age by the time we invested in an incubator, and two of them began laying beautiful, pastel green eggs. We added several of their eggs to the dark and light brown, speckled, and cream colored eggs the older girls were laying, and turned on the incubator for its inaugural run. It looked like a box full of Easter eggs.

Back and forth the incubator rocked, mimicking the hen's instinct to rotate the eggs. The incubator's temperature was set at a steady 99F, or 30C, and equipped with a fan and a little trough of water below the eggs to keep humidity level just right. On day nineteen, two days before they were expected to hatch, we took the incubator off its rotating base, initiating lock down mode; the countdown was on.

The kids and I were beside ourselves in anticipation. The seconds ticked off at a tortoise's pace until hatching day finally arrived, and we could hear a faint peep, peep, peep from inside the shells. Soon, several eggs had small cracks in them. Then, one by one, the top of the shell would pop off, and an exhausted, gangly chick would flail out of its cramped quarters. Within the hour, its feathers would dry off and fluff up. It would stumble across the top of the other eggs still in the process of hatching, clumsily stretching and experimenting with its newly discovered legs.

We often invited people over to come visit our newest, fluffy farm friends, and many of these chicks ended up as our kids' show-and-tell. They needed to bring two rhyming items? No problem—they brought a stick and a chick. They needed to bring something to represent Spring? Perfect, the chicks hatched in Spring. We figured out many creative ways to get those chicks into their classes.

The year that Rae-Rae was in kindergarten, we brought the incubator to school, and let the chicks hatch in her classroom. My daughter and I instructed the kids on how to candle eggs by holding a light up to the bottom of the egg to determine if a chick was growing inside. Most of the time, the developing chicks looked like a dark blob, a little larger than the yolk; occasionally, red blood vessels could be seen cascading down the inside walls of the shell. The veins looked electric when the light shone on them.

Each batch of babies spent their first couple of weeks living in our laundry room. Snug in a plastic box filled with shavings and an infrared heater, the chicks would learn to eat and drink from our four little human mother hens. Our sprouts would take turns tapping on the edge of the water bowl and food dish, mimicking a mother hen pecking to teach her babies how to eat and drink. Once the smell was too much to handle indoors, we would move them, and their heater, into a giant box in the garage to keep them cozy. Within days, they would transform from cute little fluff balls to awkward teenage chicks, half covered in their silky down, the other half, their adult feathers. They would practice flying up to the top of the heater or water dish, and whenever I would reach in to give them food or water, they would charge my hand. The only thing keeping the rebellious teens in, and our curious farm cat out, was to keep a wire fence panel over top of the box. By this point, they had practically outgrown their space. There was still room for them to stretch their legs, and flap their wings, but they wanted more; they wanted to scratch and peck in the dirt and fly from one object to another. The cardboard box, filled with shavings and manure, would soon head off to the compost pile to feed the worms. The chicks would soon move to their outdoor chick brooder.

The friends and acquaintances that came to visit the chicks quickly caught on that we were selling our pasture raised eggs. We began to have a steady stream of customers stopping by the farm. The first time a complete stranger came to pick up eggs, we knew that we were onto something. We realized that if we wanted, we could move from a handful of chickens to a backyard chicken operation to accommodate the demand. The thirty-five heritage birds we added to our little flock were a dual purpose breed called the New Hampshire. Although they do not lay as many eggs as commercial chicken breeds, their ability to survive with minimal inputs made them a great choice. The eggs we got from those hens were big and brown; and it was pretty common to get several double-yolked eggs a day from them. Plus, when they were past their laying prime, they would make an incredibly flavorful, nutritious bone broth.

A closeup of a freshly hatched chick. I quickly showed my sprouts before placing it back inside the warm incubator to dry off.

Our first hands-on experience with butchering came on a warm evening in early fall, just over a year into keeping chickens. Mr. Green Thumb, the girls, and I were all walking through the chicken run and around the orchard, checking on the health of the birds and trees in the last of the season's heat. Just under half of the chickens we had hatched that summer were cockerels, and we were raising them up to a certain size in order to butcher them. In the meantime, they had gotten big enough that we had moved them from the chick brooder into the big chicken pen to live out the rest of their days. We had kept two roosters since the previous fall, and seeing how gentle they were with the hens and our kids, we didn't think anything of the girls playing inside the run.

Without warning, one of the larger cockerels jumped up in front of our older daughter. She stumbled backward, away from him. There was no time to warn our youngest, who, facing the opposite direction, had her back to him. He jumped up on top of her, spurs erect, ready to attack. He left two blunt, red scratches down the length of her back. At this point, I do not know who was louder, the clucking of fifty confused, defensive hens, my wailing toddler, or her papa bear, who charged after that rooster, letting him know in no uncertain terms that he was destined for the soup pot. Holding him upside down by his feet, he determinedly strode to the backyard, and out of the girls' eye line, motioning for me to get a knife.

Just like that, with nothing but a cutting board and a chef's knife, the head was severed and lying in the grass; and my husband was ferociously pulling out feathers. I ran back to the house, grabbed the biggest stock pot we owned, and filled it with the hottest water our tap could possibly produce. I knew that the only way to get out the stubborn pin feathers was a soak in scalding water, and a lot of determination. The task had looked straight forward enough on YouTube, but it was a whole lot harder sitting in the grassy field at dusk, using only hot-ish water and waterlogged gloves, and without a clue as to what we were doing.

We butchered that carcass in every sense of the term. Chopping off his head was the easiest part of the job. We gave up on pulling out more than a few of the pin feathers. We sliced up the thigh and breast meat while removing the entrails. Roughly one hour after it all began, we had a pile of meat and bone; it was nothing resembling the whole roaster we were hoping for. Clearly, it was going to take some practice to get the hang of it. The next day, I made an appointment with the local chicken butcher to deal with the rest of the cockerels. It seemed the most humane way to handle the remaining birds until we had a mentor to help walk us through the process of home butchering.

Several of the hens we had hatched ourselves had become master escapers. It seemed that they only came back to the chicken run to eat and sleep. People that came to pick up eggs often parked half way up the driveway for fear of running over a chicken. They spent the majority of their days wandering around our field, selecting the most tender, juicy leaves and bugs they could find. Clover, plantain, dandelion and veggies from the garden, their diet was bursting with nutrition and chlorophyll.

The hens even made themselves a nest in the corner of the garage, in a little pile of straw. When we first found it, the nest was piled high with eggs. From then on, we started collecting from both the main coop and the garage. Maybe it was my imagination, but I always thought the yolks from those few eggs were just a little more vibrant yellow than the rest. The rest of the hens had plenty of space to explore as well. We fenced in about half an acre where we planted our orchard. As the shoots of plants and grasses pushed their way up through the dirt each spring, the chickens would wander further and further from the coop, scratching and playing in the dirt. Mr. Green Thumb scattered a wildflower seed mix in their run, and as spring went on, it became a sea of blue and purple lupins.

Once the weather became drier, the chickens discovered the

roots under the cedar tree on the edge of their run. Daily, the chickens lumbered up to the tree, scratching and pecking until they had carved out multiple holes around its roots. A few were so deep that the chickens weren't even visible from the fence. They fluffed up their feathers, and flapped their wings to cover themselves with dirt. Many of them would sit in their holes, soaking up the sun for hours.

As the seasons passed from one to the next, the color of the chicken run changed too. When summer came, the field filled with white yarrow, orange poppies, pink cosmos, and purple agastache. In late summer, the daisies and rudbeckia were in full bloom. The eggs followed suit. As the days grew longer, the egg yolks went from a rich, golden yellow to a deep, fiery orange. Everyone who ate them loved the color and flavor of the eggs as well.

In the first year of our chicken-keeping journey, we went from a few feathered friends to around fifty birds and a group of loyal customers. Over the past few years, we have averaged between fifty and ninety layers. We're still novices at keeping chickens, but we're learning a lot. We're learning which season, and age, their peak laying time is; and when to transition the older birds from the nest box to the freezer. We're learning how many birds our half acre run can support without getting too scratched up. We're learning how to keep egg production up during the winter, by adding more hens and a bit of artificial light.

We have also made mistakes. Small mistakes, like forgetting to double check that the kids closed the gate after they fed the

A collection of our mixed heritage hens and muscovy ducks, along with Mr. Peabody, our peacock.

chickens—on more than one occasion I have stepped outside to discover eighty or so hens milling around our driveway, flower beds, and garden. And big mistakes—like leaving little gaps along the bottom of the fence that seemed too small to be a problem. Because of this, many of the smaller chickens we had this past fall were able to escape into the neighbor's cow field, and were lost to coyotes. Thankfully, once we realized this was happening, we quickly filled in the spaces with scrap wood, and a few shovels full of gravel, and didn't lose any more birds.

While we weren't, and still aren't, the perfect homesteading farmers, we knew that this was the beginning of providing our family and our community with flavorful, nutrient rich, ethically raised food. We couldn't wait to see what the future of our homestead would become.

Rhubarb

"As one of the first vegetable crops in spring, it's appearance on the tables in late May or early June was always greeted with appreciation and satisfaction."

- Norma Jost Voth

A strong correlation exists between rhubarb and my grandparents. They ate rhubarb often, and my grandmother utilized the stalks in every sweet way possible—she made jams, juice, pies, crisps, and sauce. My grandma's pantry and freezer were full of her sweet rhubarb concoctions. Their rhubarb also had a prominent place in their garden. It stood vigorous and tall in the corner of their vegetable patch. The plants, much like my grandparents, were hardy and robust, representing the tenacity of many a weathered homesteader.

My grandma taught me how to harvest rhubarb properly, by gently wiggling one of the stalks until it released from the roots, making a distinct popping sound when it did so. We would continue harvesting several from each plant until she had enough for the dish she was preparing. Then she would pull out an ancient paring knife from the pocket of her apron and chop off the leaves and the ends of the stalks.

Back in the kitchen, she would often cover the bottom of a small dessert bowl with sugar and let me dip the crimson stalks into it as a snack. The mouth-puckering flavor of the tart fruit sharply contrasted against the sweetness of the sugar. As the rhubarb dipped into the bowl, the sugar slowly turned from white to rose. When my cousins visited, I would take them down to the rhubarb patch, and we would dare each other to eat a bite of rhubarb without cringing.

When we moved onto our homestead property, I was eager to start growing my own patch of rhubarb. We established the layout of our garden the summer after moving, and we decided that our rhubarb would be planted in the top, southern facing corner of the garden, out of the way of the raised beds and the other perennials we had in mind to plant.

It so happened that my brother-in-law, an organic vegetable farmer, planned on dividing his rhubarb plants that summer and offered us multiple roots. The timing was perfect. He showed up later that week, the back of his pick-up crowded with bins containing the rhubarb. In the heat of the afternoon, we dug holes for the plants; they needed to make their way back into the earth as soon as possible, and we did not delay.

Wasps buzzed among the plants as we worked. When I sliced through one of the larger plants with the tip of my shovel to divide it, I exposed a papery nest, wedged in among the stalks. Both he and I ended up with several stings, as he quickly scooped up the agitated wasps' nest in his shovel, and flung it over the fence into the neighboring cow pasture. My brother-in-law then stooped to the ground only a few steps from the truck and plucked up several leaves of a common weed. He told me it was called plantain and

An early harvest of rhubarb on the homestead.

passed me a couple, instructing me to crunch up the leaves in my hand and then rub the leaves with their juices onto the bites. The pain didn't completely disappear, but there was almost immediate relief as the cool plant rubbed against the burning welts from the wasps' bites.

We got all fifteen rhubarb plants tucked safely into the ground and heavily watered before we called it a day. That evening I did an internet search for plantain, the plant I had known until today only as a prolific weed. I marveled at this common, overlooked plant's ability to provide relief from an insect's sting or bite, or even from the burning rash of the stinging nettle plant. It was incredible that in the very place we had been bitten, we had at our fingertips a remedy for our pain.

Later that summer, my aunt brought me six baby rhubarb plants that she had dug up from around her own rhubarb bushes. This brought our total to twenty-one. My aunt called this variety strawberry rhubarb; they were a slightly later producing and more delicate tasting variety than the ones we received from my brother-in-law. They were also more significant, because they came from the very garden I had played in as a child in my grandparents' backyard. I carefully mulched and watered those plants, pampering them in hopes that they wouldn't die from the relentless heat of the August sun. Thankfully they survived that first summer; and as they appear again every spring, they become more like the vigorous plants I remember from my childhood.

Now, I watch my own children dare each other to take a bite of the tangy rhubarb stalks without letting their faces pucker; and I find myself pouring a bit of sugar into the bottom of a dessert bowl for them to dip their rhubarb stalks into. Last week, we picked our first rhubarb of the season, my girls and I. The April day was warmer than average and they were in shorts, tees, and bare feet as they walked between the rows of rhubarb plants, as tall as their waists.

I stood to the side as they plucked each stalk from the plant. I watched as they wiggled and popped each stalk off the parent plant. When they each had a good armful we walked back to the house to prepare them. I supervised while my girls washed each stem and then sliced them up and put them into the dish. They took turns slicing up a stalk at a time, impatiently waiting their turns to wield the knife. I gave instructions as they measured the

CJ and Rae-Rae helping me gather one of our later spring rhubarb harvests. They don't mind getting their hands dirty around the homestead.

basic ingredients for the crisp they were making—the crisp that my oldest daughter would share with her class the following day for her presentation on her family's heritage.

I was surprised when my daughter told me that only three people from her class, including her teacher, had tasted rhubarb crisp before. It was a staple I had grown up on—the epitome of a springtime dessert, and a reflection of our heritage of hard working farmers who wanted a simple dessert to put on the table at the end of a well-earned meal. It didn't surprise me when my daughter said that her classmates thought it was too sour. It is an acquired taste in a world of sickeningly sweet desserts and treats. I could see the pride dance in her eyes as she told me this. She had made this crisp for her peers almost entirely by herself, and she had enjoyed watching them squirm and grimace as they bit into the tart dessert that she had enjoyed her whole life.

This past winter on the west coast was a difficult one. We had more snow on the ground than we have had in the past decade, and the freezing temperatures continued on much longer than is common. So in March when I saw the tiny chartreuse leaves of the rhubarb plant pushing up from under the mulch, my heart did a little flip-flop in my chest. It wasn't just in anticipation of harvesting the crimson red stalks, although that was certainly something I was looking forward to.

Seeing those leaves was a sign that spring was finally on its way. It was as if winter itself was finally conceding, making way for a new beginning, the season of hope and renewal. It was also a reminder of my grandma's garden, and for a second in the middle of the rhubarb patch, I felt a pang of sadness because I could no longer share moments like these with her. But I also felt a sense of contentment, knowing that I could give my own children the gift of these memories, the way my grandma had given them to me.

Rhubarb Crisp

Ingredients:

10-15 stalks of rhubarb (2-3 lbs.)

1 cup plus ¼ cup sugar

½ cup butter

2 cups quick rolled oats

Directions:

• Trim leaves and ends from rhubarb if it hasn't already been done.

• Rinse and slice into 1-2 inch pieces.

• Toss pieces with ¼ cup sugar.

• In a separate bowl, melt the butter, and then add to it 1 cup of sugar and the oats; stir until well combined.

• Place sugar-coated rhubarb evenly in a 9x9" pan. Place oat mixture on top and smooth out until all the rhubarb is covered.

• Bake at 350 degrees for 1 hour. Serve with whipped cream or vanilla ice cream.

Pasture Raised Pigs

> *"[P]igs have a wonderful plow on the end of their noses. In this, they are truly distinctive. This means the glory of the pig is in its ability to move things around, to till things, and disturb the soil...They are part of an intricate choreography that pulls its dance moves from millennia of porcine glory."*
>
> *- Joel Salatin*

After chickens, pigs were the next animal we decided to raise. With a little sleuthing, we learned that there was a good market in our area for locally and ethically, pasture-raised pork. Plus, we had a growing egg customer base we hoped we could tap into. We planned to fill our own freezer, and then sell the rest. I remember the night Mr. Green Thumb and I discussed the benefits of adding pigs to the farm. We talked about the heritage breeds available, each breed's character traits, and the differing flavor profile of each of them.

True to his get-it-done nature, Mr. Green Thumb called me the next day with the news that he had reserved two Ossabaw piglets from a farm on Vancouver Island. The farmer would be delivering them the next week. By this time, our chickens had outgrown the portable chicken tractor we had started with; we decided that this could be the piglets' first home. Stuffed with a mound of dry straw, and outfitted with food and water dishes, their new digs were ready.

The day they arrived was sunny and warm. The farmer parked on our driveway, in front of the chicken tractor. She strode to the back of the truck and let down the tailgate. After a quick introduction, she tugged the giant pet crate to the edge of the truck box and unlatched the door. The piglets began to squeal. So did our kids. We were gathered around her like visitors to a petting zoo, waiting for a turn to touch the animals. When the farmer reached for the piglet closest to her, he let out a wild scream and flailed around, trying to escape. Sure enough, when she got ahold of the second one, that piglet hollered and thrashed as well. I could not believe that something so tiny and cute could make such an awful noise. No number of internet searches could have prepared us for the noises those piglets made.

While our kids had named the first chickens we bought, these cute little piglets were off limits. We decided early on that any animal we were raising for meat would not be given a pet name. Instead, to keep things in perspective, we called them Bacon and Pork Chop. Not long after Bacon and Pork Chop arrived at the farm, our youngest farm girl, CJ, asked what a pork chop was. She had never eaten one, as I had not found a source for ethically raised pork before we began raising our own. Occasionally, I had

Pasture Raised Pigs

One of our pigs, appropriately named "Bacon," coming to the fence to say "Hi" and see if we brought any treats.

splurged and picked up organic, thick-cut, double-smoked bacon, but otherwise, pork was not on my grocery list.

Not ready to discuss life and death matters with my then three year old, I told her that pork chops were just like bacon, which she devoured without hesitation. For the time being she was satisfied; I on the other hand was completely relieved. Our oldest had given up eating chicken for weeks after he discovered that chicken meat came from the actual bird. I hoped my youngest's first experience with our home-grown pork would not have a similar ending.

As the piglets grew, so did the size and smell of their poop. Their time in the chicken tractor parked on our front lawn was coming to an end. We needed to find them a different location, and fast. They had turned up the grass inside the enclosure, so that all that was left were mounds of dirt, clumps of grass and trampled hay, making it hard to push the tractor anywhere using man-power.

We tied up the chicken tractor to the back of the truck and slowly dragged it toward the back field. The rope stretched and groaned under the strain of the load and the uneven ground it rolled over. We positioned it near the fence along our property line. We also fenced off a sizable pasture for the pigs to roam and root to their hearts' content. When we released the pigs into their new space, they galloped around the yard, excited to have the space to run and fresh greens to chomp on.

Multiple times a day that summer, our girls would bring fistfuls of freshly picked grasses and clover stems to the pigs to snack on. We made a mud pit in the lowest corner of the pasture for them to wallow in so that they could keep cool. On really hot days we would set up a sprinkler that sprayed over onto their side of the fence, so that they could walk through it. Our pigs were happy creatures every day they were with us.

The piglets had started out as adorable little bacon seeds; by the time they reached butchering size, they had become hairy, ugly, noisy demanding porkers. I had no trouble saying goodbye to them that final day. We had gleaned several bags of windfall

Our pork, hanging in the walk in cooler at the butcher, before getting cut and wrapped. Legally, we have to slaughter and butcher our animals at a gov't certified facility if we want to sell the meat.

apples, and had spoiled the pigs with several apples each day for their last week. Now, it was time to coax them out of their pen and into the trailer with the last few apples we had. An experienced, pastured pig farmer and friend had come to help us round them up and drive them to the abattoir, which is a fancy sounding name for slaughterhouse. Since it had been raining that day and the several before it, he had to park at the edge of the driveway, not wanting to get his trailer wheels stuck in the soggy grass. That left about a two hundred foot stretch between where the pigs were, and where they needed to go.

In theory, when pigs are hungry, they follow the food right up the ramp and into the trailer. Since they had not eaten since the prior evening, we figured it would be easy enough to motivate them with apples and a bucket of grain. Apparently, this was not enticing enough, and the cabbages and greens in my garden were too tempting to pass on. Thankfully we were already into October, and there was not much that could be damaged. While Mr. Green Thumb and our friend weighed more than the pigs, it was not by much; they definitely had their work cut out for them. It was quite humorous, as a spectator, to watch two burly men chasing two stubborn porkers. I kept yelling at Mr. Green Thumb from the sidelines, imploring him not to step in and compact my garden beds. The pigs paid no attention and trampled wherever they pleased. The next spring, I remember finding some fava beans sprouting up out of several large boot prints in the raised beds.

Catching their breath and strategizing, the men grabbed a sixteen foot fence panel from under the back deck. They slowly approached the first pig, which had also settled down and was rooting up one of the overwintering broccoli plants. Before the pig knew it, they had it corralled. They slowly led it to the truck, and up the ramp. Then they walked up to the second pig, which more willingly complied. Once the guys exchanged pats on the back, and exclamations of good job and we got 'er done, our friend jumped

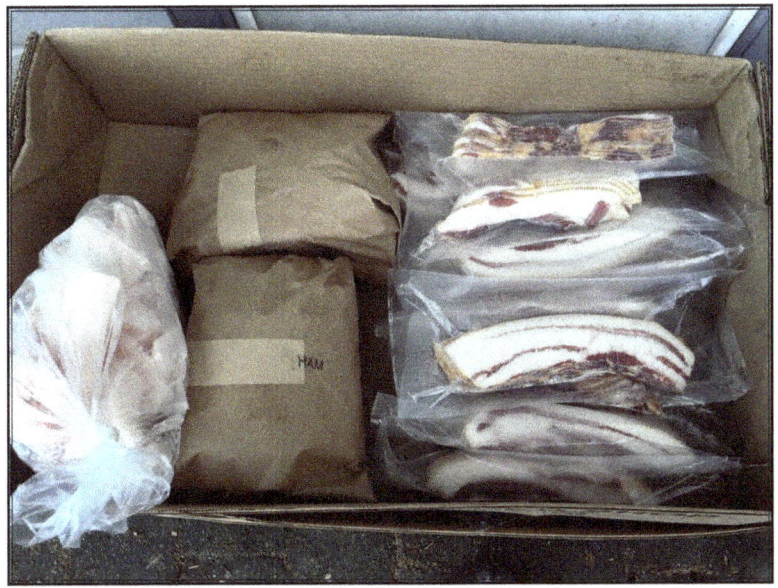

Some happy customers will be enjoying this box of various pork cuts.

into his beat-up pickup, and pulled the trailer off our property and onto the road, ready to carry the pigs to the slaughterhouse where they would also be butchered and packaged. Our first experience raising pigs had come to an end. Pork Chop and Bacon would soon return to us as wholesome food to sustain and nourish our family and friends.

I clearly remember the day the meat arrived back to us, wrapped in little brown parcels. Farmer sausage, ham, bacon, pork chops, ground pork, and breakfast sausages. And of course, several bags full of creamy-white lard. I felt as excited that afternoon as I had on Christmas morning as a child. This was one of the most exciting packages I received that year.

Maybe it was because it reminded me of the stories my mom had told me when I was younger, about the pig butchering day that they hosted at their homestead every year when my mom was just a child. Early in the morning, cars full of aunts, uncles, and cousins would pull up to the house. The men and older boys got straight to business, and took their rifles out to the pasture near the barn where the pigs were kept. Meanwhile, the women began cooking up a storm in the kitchen, getting lunch together early, so that they would be ready to begin processing the meat when it arrived.

Soon enough, the first of several broken down carcasses was delivered to the outdoor kitchen. Several ladies took turns at the meat grinder, while others mixed the meat with seasonings, and still others cleaned out the intestines and then slipped them onto the sausage maker to be used as casings. A big ham hock was added to the pot of beans simmering on the back of the stove. A large pot held a pig's head, which would be made into head cheese. The feet were scrubbed clean and went into canning jars to be pickled. A vat rendered the lard on an outdoor cooking fire, and my mom remembers fishing out and nibbling on the little pieces of cooked meat, called cracklings, from the pot as she stirred it.

At the end of a long, exhausting day, sausages, hams, pork belly, and lard were all divided up between the families. Cousins hugged, and aunts kissed the children goodbye. From here, all of the meat was either cured in the cellar or smoked in the smoke house in order to preserve it. My mom's family certainly knew how to eat "nose to tail." Not a thing was wasted.

Giddy at the thought of eating the healthiest pork available, which we had raised ourselves, I turned on the oven and stuffed

One of our pigs, covered in mud from her earlier mud bath, chomping down on a pile of grass and weeds.

two farmer sausage rings inside. Supper was not for another couple of hours, but I couldn't wait. I needed to know what our home-grown pork tasted like. I was not disappointed. We rounded up the kids from all corners of the house, and gathered them at the table for our monumental meal. We had been eating our own pasture raised eggs for over a year, but somehow this felt more significant. At that moment, I felt like a true homesteader. I had sliced up the rings into coin sized pieces, and the kids grabbed them off the serving platter, equally curious to try it.

Grateful for the life they gave, I solemnly took a bite of the sausage. The juices burst in my mouth, tickling my tongue. Sweet, salty, gamey but mild, smoky, rich, tender—all of these flavors and textures washed over my palette. I may have let out an audible moan. While the boys dug in, the girls were somewhat hesitant about eating meat they had known as an animal. We did not rush them in their decision to partake. Our youngest reached for one first. With one bite, she decided that animal or not, it was too delicious not to eat. The boys licked their lips, noisily letting us all know how much they were enjoying it.

Our oldest daughter, Rae-Rae, reached for one then, convinced by her siblings' delight. She wanted to eat the sausage, but with one bite, her gag reflex forced it up, and she rushed over to the garbage to get it out. She sat back in her seat and shuddered, leaning over to her daddy for a hug, and the affirmation that it was okay to have reacted the way she did.

The excitement of the moment had passed; the reality of one animal giving its life for another weighed heavy on all of us around the table. Many times since then, our daughter has asked us if the meat I set on the table at dinner is from one of our pigs or chickens—she still eats it if it is. But her question serves as a reminder to remember where our meat comes from, and not to take it for granted.

We have raised several batches of heritage pigs since that first time. Much of the meat has stuffed our freezers to overflowing to last us through each winter. Several of our egg customers have bought sampler packs of our meats. However, we found many new customers as we began to advertise our pasture-raised pork for sale around town. And it appears the demand for ethically,

One of our pigs in an area of their paddock where they have rooted up all of the grass. Once the grass is all gone, we daily pick grass and clover from elsewhere on the homestead and feed it to them.

healthfully raised pork and other meat products is steadily increasing.

Our family and our customers have feasted on juicy hams for Easter, Thanksgiving, and Christmas; we have served brunch with our crisp, smoky bacon. We have shared pork burgers with friends in the backyard, fried fritters in lard for a delicious dessert, and honored special guests with succulent pork tenderloin. No doubt food has the power to bring people together; but never do we feel it with such intensity as when we sit down to a table where our pork is at the center.

Wild Spring Tonic

"Teeming in the countryside, the world over, are medicinal herbs and edible plants."

- Juliet de Bairacli Levy

The first sight of dandelion leaf whorls in the grass in early spring come as a delight after a long season of cold and hibernation. The emergence of these leaves is a reminder that warmer weather is on its way; they also point to more wild edibles and medicinals which will be bursting up from the ground again soon — for it is the dandelion that announces these wild plants' return, and it does not disappoint. Shortly after its appearance, the return of plantain can be found in the grass, and the nettles appear in a shaded corner of my garden, as they do along train tracks and roadways; their delicate leaves unfurl and stretch up through the cold air to reach for the sun.

Not only are these promises of spring pleasing to the eye; they are also abundant in nutrients. There is no better time than spring to nourish the body with the gentle healing herbs that grow in abundance in the most humble patches of dirt. After a long winter of hearty comfort foods, early spring brings us cleansing herbs to strengthen and support the purifying and digestive systems of the body.

One of the easiest ways to consume herbs is in a tisane, or tea. Tea is brewed using the caffeinated tea leaves of the tea plant; tisanes use herbs to brew an equally enjoyable cup, minus the caffeine, and often with added health benefits. The heat from the water draws out all of the water soluble nutrients, and provides a warm, comforting beverage to sip on a chilly morning, or savor on a cold evening. Drinking herbal tisanes, whether store bought or homemade, is a nightly ritual on our homestead. They may be consumed for their health benefits, or simply for their warmth and flavor; they are entirely satisfying either way.

Before our sprouts ever tasted juice, they took delight in sipping lukewarm mint, lemon balm, or chamomile tisanes, sweetened with a hint of honey and milk. They still enjoy the comforting warmth that this soothing beverage can bring to the mind and body, and often can be found brewing their own tisanes in the kitchen during the cool seasons of the year. On especially cold mornings, or if our sprouts are getting over a cold or cough, I'll send them to school with a thermos full to help get them through their school day.

Folk herbalism says that our bodies are attracted to and even crave the herbs that we need. Without fail, I find myself craving licorice tea throughout the darkest days of winter and early spring —I love the flavor, but I also know that licorice supports the adrenal glands, provides relief for stomach aches, and is soothing for coughs —all things that my body could use a little help with. My youngest, CJ, has recently started asking for the tea with the flowers, referring to chamomile. She is a very extroverted, expressive person, and I imagine that the soothing calmness that chamomile provides her little body is a great relief at times. Tulsi, a type of

Wild Spring Tonic

A cup of my spring tonic tisane, great for cool spring days.

basil, and also referred to as holy basil, is another fabulous tisane I crave during periods of heightened stress; I find that within an hour of consuming a tulsi tisane, my mind and body have settled, and I find myself able to feel calm and focused once again.

Each season brings with it a unique crop of wild medicinals. Spring, along with dandelion, nettle, and plantain leaves also brings elderflowers, cherry blossoms, willow bark, and wild cherry bark; summer yields rose petals, elderberries, wild strawberry and blackberry leaves, the leaves from fireweed, and mullein flowers and leaves to name a few. Autumn offers rosehips, as well as the roots of dandelion, chicory, burdock, and blackberry. Whether the reason for consuming tisanes is for the health benefits, or simply for the comfort that a warm mug brings, I have discovered the joy of exploring the herbal medicinals that grow at my doorstep.

Grow. Cook. Eat. Share.

Wild Spring Tonic Tisane

To prepare the tisane:

1 cup fresh, or ½ cup dried of the following:

Stinging nettle leaves

Broad-leaf or narrow-leaf plantain leaves, or a combination of both

Dandelion leaves

1 cup fresh, or ½ cup dried mint leaves (optional, but improves flavor)

If the leaves are fresh, use what is needed fresh, and then roughly chop and dry the rest for future use by spreading out on a baking sheet(s), covering with parchment paper, and setting in a warm, dry place in the house for several days.

To steep the tisane:

- Place ¼ cup dried, or ½ cup fresh mixture of leaves in a quart jar and fill with boiling water. Lightly place the lid on the jar and allow to steep for at least 10 minutes. Strain if desired. Drink this warm or cooled, throughout the day.

- Alternately, place ½-1 teaspoon of each dried herb into a metal or cloth tea strainer, and pour one cup of boiling water over it. Cover the cup and allow to steep for ten minutes.

- Once cooled, add honey or milk to sweeten if desired. Drink one to several cups throughout the day. The tisane will keep in the fridge for three days. Drink daily for several weeks.

Wild Spring Tonic

Three bunches of mint hang from the hearth over the fireplace, drying out before they are prepared for long-term storage.

My First Kidding

"Animals are such agreeable friends—
they ask no questions; they pass no criticisms."

– George Eliot

Kidding season came later to our farm than to others this year. Pictures of the sweetest little doelings and bucklings started to trickle into my Facebook newsfeed in early March. Our first time mama-to-be, Clover, wasn't due for a couple of months, and her friends Poppy and Violet were due a month after her. I longingly admired the sweet images and tried to patiently wait for the day I could let our farm friends learn of our own good news.

April, however, was a kidding frenzy. Daily, both farmers I knew from the Fraser Valley, and farmers across the country, were sharing pictures of their twin, triplet, and even quadruplet goat kids. It was hard not to feel a twinge of goat envy, knowing that I still had to wait until the end of May to meet Clover's beautiful babies and be initiated as a goat midwife.

One week into May, Clover started to show signs of early labor. I quickly gathered the birthing materials recommended by several seasoned goat owners, and made my way down to the goat barn. Although she was earlier than her due date, it wasn't completely out of the question for Clover to be in labor, and I didn't want to miss out on her big day. As that first day turned to night however, I realized that maybe she wasn't ready quite yet.

When I returned to her the next morning, she continued to show signs of labor—continuously twitching her tail; sitting down and then standing up again repeatedly; calling out much more than was normal for her; pacing around the maternity pen that I had secured her in the day before. Having given birth four times myself, I empathized with her as I watched her restless actions. I imagined the discomfort she must be feeling.

And then, she began to push. Just a little at first, but enough that I could see a little bulge of what I thought was the sack that her kid was encased in. From all the YouTube videos I had watched and articles I had read, I felt I had a fairly good idea of what to expect. But I was not prepared at all for what actually happened. Instead of straining to push out a goat kid, she was actually struggling with a vaginal prolapse.

Now, let me interrupt this story for a moment to say that I know a few of you just cringed when you read the term "vaginal prolapse." However, once you have animals, no topic of conversation relating to your animals is off limits. Our children learned a lot more about female anatomy from our goat's birthing process than they would have from a lesson on the birds and the bees. So, please forgive me for giving all the nitty gritty details. This is as much a part of the farm life we live and love, as are the adorable

Poppy, a patient mother, allowing one of her kids to ride on her back while she grazes through the field.

pictures of baby animals I post on Facebook.

At this point, I went into panic mode and took to Facebook myself to ask my more experienced goat colleagues what on earth I should do. Several goat farmers suggested I call a livestock vet, and so that is what I did. He was incredibly helpful over the phone, and was able to talk me down from my frantic worry—until I asked if I could bring her in for a C-section if she needed it, to which he replied that they weren't set up to do that type of procedure. It felt like I was back to square one.

While Clover continued to grunt and strain, she was more relaxed when I was in the pen with her. So, in my muddy Carhartt overalls, I snuggled up next to her in the straw bedding, stroking her head with one hand, while I navigated my next steps on my smart phone with the other. Thankfully, a goat farmer in the next city reached out to me, and said that if I called her, she would talk me through what she would do. I jumped at the chance.

I felt so much more at ease after I talked to her, and on her recommendation I called a second livestock vet, with whom she had personally had good experiences. Upon answering, the secretary took down all of my information, and very graciously complimented me on my knowledge as a first time goat midwife. Again, my mind was put at ease as I now waited for the veterinarian to call me back. Shortly after hanging up the phone, I received the call from the doctor that I was waiting for. After reviewing the information I had left for him, he decided against a C-section for the time being, and instead he told me I had to do the one thing I was dreading—I was going to have to push her prolapse back in.

With new resolve, trimmed fingernails, clean hands and a bottle of lubricant, I made my way back down to the barn. This was definitely going to be harder for me than it would be for her. Clover was very unimpressed with my amateur midwifery skills, and strained against my hand as I forced it inside. Not only was I attempting to push her prolapse back inside, I was also trying to feel for her cervix, and for a goat kid, to tell if she was actually in labor, like the vet had coached me to do. It took several attempts to finally push my hand in far enough to check for her cervix.

While Clover resisted, she was very patient with me trying to help her. I couldn't feel her cervix or a goat kid in the birth canal, but at least her prolapse was back in place. And my hand was only

My First Kidding

CJ holds up a goat kid so that I can check whether it is a boy or a girl.

sore for a day or two after from being squeezed from inside of my goat's lady parts.

From this point on, I checked on Clover multiple times a day. For a week, her body held everything in place where it should be. But by the following Saturday, her prolapse appeared again. Anytime she strained, even a little, a pink fleshy ball popped out. I was beside myself with worry—would she be able to give birth naturally or should she have a C-section? What if I didn't catch it in time and she or her kids didn't make it?

The next day was Mother's Day. While Mr. Green Thumb and the girls prepared me breakfast, I went down to the barn to check on Clover, and our boys finished the other morning chores. Clover seemed peaceful, and her prolapse wasn't an imminent problem. She wasn't acting like she was in labor as far as I could tell from the information I had read, and videos I had googled. Although she was now a day overdue, from her lack of laboring I felt like it would be another day or two before anything happened.

The boys and I cleaned ourselves up, and we all sat down to breakfast. We enjoyed our morning church service. We even lingered over the Mother's Day buffet that my husband had made reservations for earlier that week. Only several hours had passed since I had last checked on Clover, so when we got home, I didn't bother changing before running down to the barn to peek in on my expectant mother. There in the maternity stall was not one, but three goats—Clover and her two itty-bitty, brand new babies. At that point it didn't matter what I was wearing. I swung open the gate and went into the stall to take a closer look. The smaller brown buckskin was mostly dry, and I scooped him up to give him a little cuddle. The lighter one was still covered in birthing liquid, and I let Clover follow her instincts to clean him off before I inspected him.

They were both shivering a little, and they were still figuring out how their legs worked. They would take one or two steps before toppling over, and then strain to get back up to their feet only to fall again. Clover was so patient with them. She continued to lick them, and sniff them, and followed them around the pen so that they could find her. They instinctively searched for her teats, and she stood perfectly still, even lifting her leg to help them find

Rae-Rae and CJ, posing with the first goats to join our homestead, Clover and Acorn.

one. I quietly slipped out of the stall so as not to disturb the new mama and her precious babes.

Two healthy, adorable blue-eyed boys joined our little herd that morning. Sure, I was a little disappointed that I didn't get to be there for the actual birth. But I was also relieved that clover was able to easily labor and give birth to her twin bucklings without any complications from her earlier prolapse. Aside from my own children making me a mother, this was the best mother's day gift I had ever received.

The Potager Garden

"When the world wearies and society fails to satisfy, there is always the garden."

- Minnie Aumonier

My garden is a wild place. Out of control to some, it delights me for the organized chaos of it all. Volunteer borage tumbles from her stalks and kisses the ground; mustard stems bow low to the earth, draped across the garlic, laden with seed pods. Trellised peas, beans, and cucumbers fight against gravity and climb upwards to the warmth of the sun, depositing leaves and flowers along their vines as they grow. The perennial patch of Egyptian Walking Onions has become a mass of criss-crossed stalks, prostrate against the dark, rich soil. Their seeds are bursting from their papery parcels and will soon find themselves reborn.

The garden is sensory therapy for me. I caress the feathery carrot leaves, the delicate cilantro flowers, the dangling pea pods as I walk past; I inhale the aroma of anise hyssop, catmint, and lavender as I brush against them. I stare into the center of a sunflower stalk preparing to reveal its glorious flower, and I discover a creamy white spider weaving its web across one of its leaves. I feel the warm sun on my back as I bend to pluck weeds from between the plants; I feel the tickle of a drop of sweat, tracing its way down the small of my back. And then a refreshing breeze rushes up to greet me, and my flushed cheeks cool, and I hear the air move through the trees behind me and rustle their leaves before blowing past them as well.

The garden is a place that nourishes mind as well as body. When I step into the magic of the garden, and I recognize the small role I have in this life and death cycle, in the planting of seeds and seedlings, amending the earth, and adding some water here and there, I remember just how tiny I am, and how vast and grand the world really is. I cannot help but marvel that the sun's warming rays reach every garden on the entire globe, and that the rain falls onto each patch of tended earth in its own time. Breathing deeply of the wildness of my garden, I feel free from the stress of my day, of the world around me, of my own worries and struggles. In the garden, I find a place to just be.

I have learned a few things over the last decade and a half of gardening, and slowly, I am incorporating what I learn, translating it from head knowledge to practical, intuitive skills along the way. For instance, some plants like each other, and some do not. I learned this the hard way, when after one summer of poor growth and production with both our beans and our leeks, I googled what

The first garden beds we built on our homestead. *Photo by Nathan Jantzen.*

From the top of our orchard, looking down the hill to the chicken coop. The blackberry bushes are starting to get unruly. Our goats will be moving into the orchard soon to eat the blackberry bushes and other weeds.

would cause them to be stunted and produce poorly, as we have had success with these vegetables in the past in very similar conditions. It turns out that beans do not like growing around leeks—something I was sure to make a note of for future reference.

It made me curious though. What other annuals do we grow in the garden that have preferences or hostility for their neighbors? It turns out, there are quite a few. Beans don't play well with anything in the onion family, but they do well with corn, and its stalks provide a climbing medium to support the beans. Also growing well near beans, Summer Savory, an aromatic annual herb, happens to taste heavenly with them. Carrots do well near most vegetables, but do especially well planted around onions and garlic, as the smell from these alliums tends to prevent certain bugs that enjoy feasting on these root vegetables.

Cucumbers don't do well around potatoes but don't mind the company of lettuce, and also provide some much needed shade for the cool-loving crop, which would otherwise go to seed prematurely in the heat of the summer; just be sure to keep the lettuce away from broccoli. Tomatoes, peppers, and basil are a happy trio, and taste equally lovely together; but they dislike growing near vegetables from the brassica family. Radishes can easily be seeded around the base of tomatoes, or in between rows of spinach, but don't plant them near kohlrabi. Onions can tolerate most vegetable plants, except for beans, peas, and sage.

With this knowledge, I carefully plan the garden layout with Mr. Green Thumb's help. We consider which vegetables grow well together, as well as where certain vegetables were planted the year before. Each plant gives or takes certain nutrients from the soil, and knowing what grew there last year helps us know what to

Our cucumber plants are very happy, trellised inside our unheated greenhouse.

plant this year. Beans and peas leave a lot of nitrogen in the soil, and so I plan to avoid those areas for planting potatoes the following year, as the nitrogen produces a bounty of foliage but a less than average crop of potato tubers. I will plant carrots and beets where the beans have been, as well as brassicas, as these rely on the extra nitrogen to kickstart their growing season. And just like the potatoes, I will also avoid planting peas and beans in the same location as they were planted last year, as the nitrogen will reduce yields.

Grow. Cook. Eat. Share.

The edge of the garden is only a dozen or so steps from our back door. It's become quite natural to leave a pan over the stove to run out and grab a handful of herbs to season a dish; sending one of the sprouts out to gather a bowlful of lettuce for a salad, or green beans for a side dish, is common-place. Once the garlic, onions, and potatoes have been harvested and cured, they are stored in our garage, just beyond our mudroom. Between the freezers full of meat and our garden full of the freshest produce and fruit, we find a great deal of abundance surrounding us.

Potager, when translated, means "for the soup pot," and that is what my garden is—a means to feed and nourish. Herbs and vegetables, perennials and annuals, cultivated and wild, all mingle together to create a vibrant display of color, texture, and flavor. Stalks of asparagus are ready a few at a time, as are a basket full of tender nettle tops or a bouquet of kale or mustard leaves. This is how we eat, especially in spring. A little of many different satisfying vegetables put together create a humble weekday meal or a charmingly rustic spread for guests on weekends.

Once spring is about to give way to summer, our garden begins to flourish. We may feast on the same fruit or vegetable for weeks on end, but oh, how we feast. When the memories of fresh corn on the cob dripping with melted butter have had us salivating for the last ten months, there aren't too many complaints when corn cobs make their way onto the dinner table several nights per week. Variations of berries with ice cream and in cobblers or crisps are a regular occurrence. During the peak of the garden's performance, any meat in our meals takes a back seat to the vegetables we enjoy, harvested at perfection.

The Potager Garden

CJ is helping me harvest the potatoes. She loves to help in the garden, and follows me wherever I go.

A beautiful bouquet of freshly harvested French Breakfast radishes.

All six of us have been grazing on strawberries this past month. We pass the strawberry patch as we check on the chicks and ducklings, and when we walk to the high tunnel to water the tomatoes and peppers. We go a little out of our way to snack on a couple of berries before feeding the pigs or when we check on the lambs' water. The dazzling crimson berries stand out against their leafy green foliage. They are practically begging to be eaten, dangling from their delicate stems like succulent jewels. And just as the strawberries are winding to a close, the blueberries and raspberries begin to ripen, and the clusters of green blackberries are abundant where the goats haven't gotten to them, promising another bumper crop.

Planting a garden is an act of faith, or as Audrey Hepburn said, "to plant a garden is to believe in tomorrow." It is arguably more challenging for me than my sprouts to spend days on end watering a bare patch of soil, wondering all the while if the seeds will sprout. I think this each time I bury a seed into the soil; occasionally my worries are validated as either the seeds or the condition of the soil or weather is not right to bring about a lush crop of seedlings. Most of the time, however, the seeds sprout to reveal a new life. I feel a mixture of relief and satisfaction as each tender plant pushes its way up through the ground and into the visible world.

The Potager Garden

CJ holding the biggest head of cauliflower we've ever grown.

A box of mixed peppers, tomatoes and zucchini, heading off to a happy customer.

Feed the Bees

"The hum of the bees is the voice of the garden."

-Elizabeth Lawrence

On one of the first dry days in early spring, my beekeeping mentor and I met at her hives, which are situated on the highest point of our southeastern facing orchard. I had recently completed a provincially certified beekeeping course, and this meeting was the first step in handing over the reins of her small-scale honey operation —equipment, bees and all—to my care; she no longer had time in her schedule to dedicate to beekeeping, and I was excited to take over her colony and the heap-load of equipment she was selling at a fraction of the price for new.

We got right down to business and carefully lifted off the outer cover of the hive. She had checked on the bees just a few weeks earlier, and she had observed the worker bees busily cleaning the hive, feeding larvae, and keeping them warm. She had sprinkled some dry sugar on top of the inner cover as an extra precaution in case they needed any extra calories to get through the last remaining weeks before the flowers began to bloom and they could begin collecting pollen.

Both excitement and worry had clamored for space in my mind since I had begun to anticipate this moment. Excitement at the hope of a colony that had endured the last few weeks of a harsh winter and was now ready to take advantage of the early forage just beginning to blossom; worry that the bees did not fare as well as hoped, and were struggling to survive. She handed me the hive tool, and I pried up the corner of the inner cover, still covered in sugar and sealed up with propolis, off of the brood box, exposing the frames. It was worse than we feared; all of the bees in the hive had perished.

Since having bees on our property, and anticipating keeping my own bees, I have looked at the early spring blossoms with fresh eyes. Never before had I searched out the flowers of each tree and shrub with such fervor, as I have these past several springs. I have kept a mental note of which flowers bloom when, and this year, I've been logging dates and locations of these blossoms in a journal as well, to keep track of the seasons when my foraging bee friends have access to the sweet nectar that they convert into honey, and to the protein-packed pollen that the worker bees use to feed the larvae.

Early spring provides the first meal for the bees, who have subsisted throughout the winter on their own liquid gold, hon-

A honey bee gathering nectar from a wonderfully aromatic David Austin rose.

ey. Dandelions, hazelnuts, maples, and willows are the bees' first spring foods, and they exist in abundance in the wild and on our homestead as well. A combination of cultivated and wild flowers rounds out their diet as crocuses, rosemary, oregon grape, and the kale, brussel sprouts, and broccoli begin to bloom. Cherry, peach, plum, and crabapple trees follow on their heels, with rhododendrons and lilac right behind them. As spring truly sets in, elderflowers, apples, pears, quince, and dogwood bare their beautiful floral array; while blueberries and strawberries present a more subtle display.

Every spring I ask Mr. Green Thumb to hold off a few weeks on mowing down the dandelion patch, and every year he happily complies. He knows just as well as I that honey bees as well as the wild bee population need all the help they can get these days. Considering that all pollinators are facing real threats, and at least twenty-five percent of North American bumble bees are at risk of extinction, I want to do my part. I know that the wild and cultivated flowers we have on our homestead not only benefit honey bees, but all of the other species of bees that make our homestead their home. Not to mention that once the bees have sufficient flowers to harvest from, I harvest the dandelion flowers myself, to make dandelion infused oils for salves, dandelion syrup, and even dandelion mead.

My friend and I began cleaning out the dead bees from the boxes and assessing the frames to see if they could be reused. Between my book knowledge and my friend's experience, we determined that the bees had sadly died of starvation. It wasn't for lack of food though—as well as the sugar my friend had left as a backup, there were several frames full of capped honey comb in their hive. They had starved to death because the queen had begun laying eggs early, due to warmer than average January temperatures.

Once the eggs are laid, the worker bees cluster around the eggs, feeding them and fanning them with their wings to keep them warm. At this point, the worker bees eat the honey on the frames nearest to them and the larvae, but will not move throughout the hive looking for more once their supply is gone. If they had just hopped one or two frames over once the nearest frame

I am wafting smoke around the hive to help the bees calm down before checking each frame for egg development, pollen collection, honey production, hive health, etc. *Photo by Nathan Jantzen.*

was empty, they would have found food; but instinctually, bees won't leave their brood.

After we had cleaned up the hive, we carried on cleaning out her stored equipment. We scraped old wax off of the boxes and frames and checked to be sure no mice, wax moths, or other insects had caused any damage, determining whether the equipment was still in good enough condition to be used for another beekeeping season. We pried grungy, unusable foundations from their wooden frames by stepping on the foundation while pulling the frame in the opposite direction until it made a satisfying pop and released. We ended up with far more usable frames than expected and cleaned those up as well, before packing up all of the equipment and storing it until it would be needed.

Summer for honey bee colonies provides a combination of abundance and dearth. The lupins in our orchard meadow have been replaced with daisies, rudbeckia, cosmos, and poppies; and on the edge of the woods, the oregon grape and elderflowers have given way to mullein and foxgloves. In the garden, potatoes, squash, beans, tomatoes, and peppers provide pollen. In the lawn, the white clover provides sweet nectar, while the purple clover in the pasture holds onto its nectar until the days are above 90F, or 30C.

If our small homestead and the neighboring flowers cannot provide enough pollen and nectar for the bees, they will resort to eating honey, and the growth of the colony is stunted and even drops, as there is simply not enough food to feed the growing larvae. In the last several years, the Fraser Valley has seen earlier and longer periods of scarcity, meaning that honey bee colonies are smaller and weaker going into fall and winter, and that they may not make enough honey to feed both them and us for the months to come.

I grapple with these issues as I plan and prepare for the arrival of the bees I had ordered months ago shortly after I learned my friend's colony had not survived. Keeping honey bees is certainly not for the faint of heart, as talk of colony collapse disorder, endangered bee species, difficult pest management, and other troublesome details circle regularly in the farming and homesteading communities. I am relieved to be heading into this project aware of the challenges ahead of me, thanks to the course I took, my friend and mentor, and other beekeepers I have met and spoken

My beekeeping outfit and smoker, two necessary tools for beekeeping. *Photo by Nathan Jantzen.*

with along this journey. If nothing else, from now on, I will be planting even more intentionally to feed the bees.

My honey bees arrived at the end of May, roughly ten thousand of them, buzzing inside of a rectangular plastic box. As I donned my new-to-me bee suit, lit the shiny, new smoker, and opened the lid of my starter colony, I felt a quiet calm fall around me. It was a sacred moment as I met my colony for the first time, face to face. I gently puffed the smoke over the frames in the temporary box, signaling to the bees that it was time to eat, before lifting each frame out to examine it. In the wild, when bees encounter smoke, it is an indication of a forest fire, and their instincts are to eat as much honey as possible before the colony flees their home to find a new, safer location. The waft of smoke I waved in their direction did not hurt them; it simply distracted them from my intrusion into their home, while I quickly worked to move them to the new bee box.

One by one, I gently pried the frames apart from each other with my hive tool, and then as steadily as I was able, I lifted them straight up and out of the box. Some were full of uncapped and capped larvae, a good sign the queen was actively laying. Others had pollen and uncapped honey in the combs. One frame in particular was already heavy with honey and its sealed wax cap. On the second frame I lifted up, I spotted the queen and breathed a sigh of relief. It was my first time looking for a queen on my own, and honestly, this was one of the tasks I was most worried about being able to do solo. One by one, I carefully placed the frames into the new box, keeping them in the exact order they came in, so as not to disrupt the bees anymore than necessary. I placed the inner and outer covers on top of the box, and checked to be sure there were no smoldering ashes inside my smoker. Challenges and excitement aside, I was now, finally, a beekeeper.

My beekeeping mentor's hives at the top of our orchard, surrounded by lupine, perfect for early spring foraging.

COOK.
Summer

High Summer

"Earth is here so kind, that just tickle her with a hoe and she laughs with a harvest."

- Douglas William Jerrold

We're melting here. The delicate wisps of smoke from winter chimneys are replaced by a thick blanket of smoky smog, smothering the valley with heavy, humid air. The smoke has blown in from the north and east, from wildfires burning out of control. I say a silent prayer as I look up into the hazy sky and catch a glimpse of the glowing, orange orb perched high in the sky. The only consolation to this oppressively thick air is that the temperature will be several degrees cooler today—a gift after weeks with temperatures in the high nineties, or thirties celcius.

Every summer seems hotter than the one before. Maybe it's the effects of global warming I'm feeling. Maybe I'm hyper-sensitive to these changes now, because it directly impacts our trees and plants' abilities to produce fruit, berries, and vegetables. I had never paid attention to the weather before farming the way I do now. I certainly hadn't celebrated a sprinkle of rain in the middle of summer up until these last five years. Until recently, I never looked forward to an overcast summer's day, or week, to give our fruit trees, vegetable plants and animals some reprieve from the heat.

With this heat begins the chaos of ripening fruits and vegetables all at once. All of our planting, watering, weeding, and tending culminates in an annual explosion of exquisitely sweet tomatoes ripening on the vine, green beans precariously perched atop their ten foot tendrils; slender zucchini that becomes behemoth overnight; broccoli florets blossoming open from the center of their foot long leaves; mystery potatoes flowering in last year's compost heap.

Our freezer is bursting as I try to use up the remains of last year's produce quicker than we add to it. I sell the last of the bacon and pork chops, and render the last bit of lard to make as much room as possible for the incoming produce. We eat a Christmas ham in July and say farewell to the last of our home-raised pork until we receive our new batch back from the processor in early September. Bags upon bags of blueberries are stacked up along one side of the freezer. And then comes blackberry season. They are monstrous this year—plump, juicy, succulently sweet. I tuck a few bags of these inky-purple gems into the freezer as well.

The three-month-old kittens—we have five in the house at the moment—are scattered around the house, sprawled out in a semi-permanent napping position or relaxing in front of the fan I have positioned towards myself in the kitchen. During the heat of

A variety of heirloom tomatoes, cherry tomatoes, sweet and hot peppers, and miniature eggplants.

the day, my kiddos are also often napping in front of a fan or in the cool of the basement.

The pigs are attempting to stay cool by taking turns lounging in their water trough or rolling around in the slippy, gloopy mud hole. Since pigs don't sweat, they need the protection of the mud to keep them cool. But beware when they shake their bodies off after getting out of the mud.

One of our Muscovy hens, who we assumed had flown the coop last month, came wandering out from the marshy weeds of the chicken run with five little ducklings in tow. Their peep-peeps can be heard at quite a distance as they toddle after their mama among the cool of the weeds. The goats take shelter in the shadow of the barn, or along the edge of the fence where they can find a smidgen of shade. They lie outstretched on the compacted grass, sometimes panting, other times subdued, calmly chewing their cud.

Our last doe kidded just two weeks ago; it was a little later than we had intended this year, but animals don't always cooperate with our schedules. Her triplets, two doelings and a buckling, bounce around beside her and even on top of her, as she snoozes up against the cool foundation of the barn. The heat doesn't seem to bother the doelings and buckling as they hop around through the grass in their pasture, playing tag, and head butting each other and their mama's udder when they want a drink to cool them down before bounding off to play again.

Life is flourishing all around us, plants and animals alike. And we are enjoying the bountiful harvest that life provides. In the kitchen, crocks and jars line the counters, full of cabbage, green beans, carrots, and cucumbers, all fermenting in a salt brine, which become tart and tangy after about a week spent steeping in this brew. Vats of fruit mixed with cane sugar are bubbling away on the stove, getting prepared to be turned into jam. Most of the tomatoes that we harvest on a daily basis are tucked into freezer bags and popped into the deep freeze.

It's hard to remember that this abundance will be over soon; sometimes I wish these days would come to an end before their time as I labor over a pot of salsa or another pot of jam well past midnight. In the moment, I think to myself that I should just buy my pickles and jelly next year. But we all know what happens when summer finally rolls around. I find myself, at my own will, standing over the stove for days and even weeks on end once again. I insist every year that I need to can my own produce so that I am in control of how it is grown, and what ingredients go into the final product.

The end results of this labor of love are incredibly rewarding.

High Summer

I ferment many of the hot peppers we grow, preserving them throughout the winter for use in dips and sandwiches, or to eat straight out of the jar.

Row upon row of shiny mason jars are lined up and stuffed to the brim with all sorts of summer goodies. Occasionally I open this cupboard just to admire the symphony of small, medium, and large jars arranged in a rainbow of colors. This will last us well through the darkest days of winter.

Our little air conditioner, which is propped up in our bedroom window and held in place with a few screws and a lot of duck-tape, is working harder than I am these days. Every evening our littlest sprouts fall asleep in our bedroom—the only cool room above ground. And every night before bed, my husband and I carry them to their own beds. I feel a twinge of guilt as we carry them out, knowing that we get to relax for the entire night in the comfort of our temperature-controlled room, but I push these feelings aside; I know that children can sleep through almost anything, and the fact that I couldn't possibly get a good night's sleep without it.

I can hardly get a decent night's sleep as it is, between the pulsating hum of the air conditioner and the continual loop my mind runs of all of the jobs that need to be done in the coming days and weeks. Making a list only helps a little, as I am constantly remembering more things to add to the list, which happens most often in the middle of the night. Jotting down a quick note in my smart phone alleviates the stress of having to remember, but my brain often cannot shut off again for quite some time. Harvest the peaches, it says; check the ripeness of the crabapples, it demands; check on the animals' water, and feel for horn buds on the goat kids; it retorts—and the list goes on.

We've managed to carve out a couple of weekends at the ocean the last two summers, thanks to a friend of a friend who feels comfortable caring for our animals. She grew up on a farm herself and is well versed in the pigness of pigs and the chickenness of chickens, as Joel Salatin would say. More importantly than

High Summer

The cucumbers are ready to be processed and turned into pickles.

that, I feel confident leaving our animals in her care. And I know that I—and the vet—are only a call away, should anything arise.

As a family, we have basked in our weekends of much needed rest to reenergize us for the busy months on our farm; and the time our family has spent together in tight quarters, spanning preschooler to teen, has been unusually pleasant. Although I must admit that most of the time we spend together during daylight hours happens on a boat. We often joke and laugh together in a way that rarely otherwise happens, as we find ourselves bobbing along with the current and the waves. Our sprouts take pride in dropping and reeling in the crab and prawn traps, and catching lingcod, rockfish, and the occasional sockeye.

On the boat, our boys especially, are important contributors to the seafood that fills our bellies that weekend, and our freezer at home. They are both seasoned pros at baiting the hooks, jigging for fish, reeling them in and netting them. And our girls are eager to learn from their big brothers. Often, Mr. Green Thumb and I have sat back and watched the older teach the younger, as we once taught them. The patience they both exude in helping their little sisters in moments like this is precious to watch.

Back at the cabin, the girls take charge, carrying a bucket full of crabs over to the tree stump in the yard we have dubbed, the altar. Our girls bravely reach into the bucket and pull out each crab by a leg, being careful to keep their fingers far away from their clenching claws. As they lay each crab down on the altar, they examine the underside once again, confirming that it is indeed a boy crab and not a girl, by checking the design on its abdomen.

Then, almost gleefully, one of them will plunge a blunt stick into the rounded part of the abdomen and pierce through its flesh and shell, killing it swiftly. I can't remember who came up with it, but years ago, one of our sprouts lifted up an impaled crab and declared it a crab popsicle—this ritual has been repeated ever since. It's an assembly line after that: pop the shell off, break it in half, pluck out the gills, rinse and repeat. From here they go into the fridge until we have amassed a heaping plate full of crab carcasses that we feast on the last night we are there.

After breakfast most days, our little sprouts beg their older brothers to walk them down to the dock and jig for fish. Secretly, the boys enjoy it although they would never admit that to their little sisters. The begrudgingly collect their tackle boxes, rods, and life jackets and catch up with the girls who are already half way down the path to the dock. Upon return, the four of them are eager to relay the morning's successes. How many fish each one

Rae-Rae is brave, and helps us kill and clean the crabs we catch on our weekends away throughout the summer.

caught, who caught the biggest, if the girls baited and cast the lines themselves. The boys applaud their sisters' efforts, and the girls glow in their older brothers' approval.

We return home after these weekends, refreshed and ready to tackle the days and weeks ahead with a hop in our step. I feel giddy as I unpack the fish fillets and yogurt container tubs full of BC spot prawns into the freezer. Not even minutes after we have arrived and unloaded the truck, our four sprouts return to the routine of teasing, pestering, and bickering with each other. The spell of the salty ocean air has been broken and we are officially back to reality; back to the vegetables that have multiplied in the three days we have been away; back to early mornings and late nights; back to the list of chores and tasks we have yet to complete; back to the place we call home.

Farm Fresh Eggs

"A box without hinges, key, or lid,
Yet golden treasure inside is hid."

-*J. R. R. Tolkien*

I usually wait until early afternoon to gather eggs from our sixty or so hens. It gives the early birds a chance to get in and out, and the stragglers a chance to finish. Today I didn't get to the coop until early evening. The sun was low enough on the horizon that the air had already cooled a bit from the intensity of the earlier summer sun.

As I stepped into the rustic coop, I felt the hot, stale air hit me in the face. There were already chickens sitting up on their perches, getting settled in for the night. There were also about half a dozen hens sitting in the nest boxes. One by one, I gently lifted them to retrieve their eggs. They weren't pleased with the disruption. Some ruffled their feathers, others pecked at me, and some jumped off the clutch of eggs they were sitting on, squawking and flapping their wings.

As I reached into one nest box and scooped up a pair of eggs into my hand, I felt a warm, squishy deposit land on my exposed toes. Even the girls up above the next boxes on their perch were defending the sitting hens. I was not amused but relieved that the hen's droppings had only landed on my foot. I finished collecting the last eggs from the last nest box, and as I turned to leave, balancing an overflowing bowl of eggs, I saw a rat skitter across the shavings on the floor of the coop.

Mr. Green Thumb had mentioned that he had seen a couple of rats the last few nights, but other than the rat floating in the ducks' water trough, I hadn't seen one in quite a while. It was unexpected, and I jumped, almost losing several dozen eggs in the process. I ran out of the coop and slammed the door with my foot as I left. I shuddered at my close encounter with that not so tiny vermin, and the lack of protection my flip flops would have provided had it run across my feet; as I walked back to the house I vowed I would collect the eggs earlier the next day.

With eggs, it seems there is never the right amount. When the hens are in full spring and summer production, they lay so many eggs that I've had to give some away, or feed them to our pigs, just to stay on top of them. During the craziest production season, many of our customers go on vacation or take a break from baking so as not to heat up their kitchens.

I have modified many recipes to incorporate more eggs; almost every meal includes them in some form. Waffle, cookie, and muffin batters hide multiple eggs in each batch. Not to mention

A collection of freshly washed blue, brown and white eggs, laid by our heritage hens.

the abundance of egg-laden desserts such as custards, meringues, and ice cream. Regularly in the summer months, we'll eat eggs in some form for the evening as well as the morning meal. Eggs are an integral part of our diet—for about six months of the year.

As the days shorten, and when the sun hangs low in the sky throughout the fall and winter, eggs are in short supply, and incredibly valuable. We often go without eggs so that I can still scrape together a couple dozen a week for our loyal customers. When I do reserve a dozen for our own weekend breakfasts, I feel indulgent. Even something as basic as a soft poached egg tastes decadent after going a week without.

Our hens have finally completed their annual molt, which seems to be timed with the darkening days of fall. Roughly once a year, the chickens go through a tiresome and rather painful process of shedding their old feathers and growing in new ones. The girls become noticeably more skittish during molting season and would much rather lie around in the nesting boxes or sit on their roosts than forage for greens and bugs. Because of the amount of calcium and energy it requires of the hens to grow new feathers, egg production drops to zero.

Molting usually happens gradually, over the course of a month or two; but a handful of birds get hit hard, and they lose the majority of their feathers all at once. At this point they look more like a drowned rat than a chicken. Pink patches of tender skin are exposed, and if they happen to be at the bottom of the pecking order, the other birds will mercilessly pick at the exposed flesh, leaving a bloody mess. Although this has rarely been the case, we have had to isolate a couple of birds over the years, by placing them into a private pet crate until their wounds heal up.

Since our girls have finished molting, we have turned a light on in their coop to stimulate egg production again, and they are responding just as we had hoped. In the weeks since the light has been on, we have watched our egg count increase from half a dozen, to one, and then two dozen eggs daily. Our girls are currently producing near our summer rate of around three dozen eggs each day. And we are scrambling to encourage our customers to come to the farm to buy our eggs, when just last month we were turning them away.

Breakfast during the summer is never dull. Fried eggs with sauteed asparagus and radishes are a delicious early summer treat.

There is nothing quite like an egg straight from a pasture raised hen. The first notable difference is the color of the eggshell. Different colored birds lay different colored eggs—white hens lay white eggs, whereas brown hens lay a variety of brown colored eggs which range from a very pale beige, to a terra cotta, to a rich dark amber color; some even have a slightly pinkish hue. And they may also have dark brown or white speckles.

There are also several heritage breeds that lay colored eggs ranging anywhere from grey to baby blue, from pastel green to olive green. The young pullets we hatched in the spring are laying now, and we are getting beautiful pastel and olive green eggs from these heritage cross birds, aptly known as Easter Eggers. Each egg carton I fill is once again a lovely assortment of white, light and dark brown, speckled, and light and dark green eggs. I never tire of seeing my egg basket full of such beautifully colored eggs.

The size of the eggs is very different from the eggs sold at the store. The size of the hen affects the size of the egg; petite breeds of hens can lay smaller than average eggs, and bigger hens can lay very large eggs. One breed we raised, called New Hampshires, regularly lay double-yolked eggs that were two times the size of eggs from a grocer's dozen. Eggs can also come in all shapes—from very narrow and pointed at one end, to an almost perfect sphere—and everything in between.

Upon cracking open the egg, several more differences are apparent. First of all, the shell is usually quite thick, and hard to crack open—this is because of the hens' healthy diet. The only complaint we've ever received is that the eggs are incredibly hard to peel once boiled, which means that deviled eggs don't look quite as attractive when made with farm fresh eggs. The second difference is seen in the color of the yolk. It is comparable to the golden-orange glow of a summer sunset. All of the chlorophyll that the hens have chomped down on, have been converted into the most vibrant yolk possible. And instead of the whites running all over the bottom of the bowl when opened, they sit tall and plump up around the yolk.

And then, there is the nutritional superiority of pasture raised eggs compared to conventional. Hens who get a daily dose of time outdoors, foraging for bugs and consuming leafy greens rich in chlorophyll, produce eggs that are much higher in vitamins A and D, and omega 3s. These eggs also contain more of the "good cholesterol" and less of the "bad cholesterol." And finally, there is the taste. Pasture raised eggs taste richer and creamier than store bought. Many of our customers who grew up on farms say that the taste of our pasture raised eggs is how they remember flavor of the eggs they ate in their childhoods. Many customers tell us that we have ruined them for store bought—which works out for both them and us.

Farm Fresh Egg and Potato Salad with a bacon vinaigrette

Ingredients for the Salad:

8 farm fresh, pasture raised eggs, hard-boiled and diced
*(see tip below)
2 lbs. new/small potatoes
6 slices cooked bacon, crumbled (reserve the bacon fat)

Ingredients for the Vinaigrette:

1/4 cup bacon fat (olive oil works well in place of bacon fat)
1/4 cup apple cider vinegar
3-4 scallions/ green onions, chopped
(roughly 1/4 cup when chopped)
1 small handful flat-leaf parsley, minced
(roughly 2 Tablespoons when minced)
1 small handful dill leaves, minced
(roughly 2 Tablespoons when minced)
1 small white or red onion, minced
(roughly 1/2 cup when minced)
2 tablespoons Dijon mustard
1 tablespoon honey or maple syrup
2 teaspoons sea salt
1 teaspoon freshly ground pepper

Instructions:

- Cut potatoes in half, and steam until fork tender.

- To get perfectly cooked hard boiled eggs with firm but not overcooked yolks, place eggs into a pot with cool water. Bring water to a boil. Once boiling, time for four minutes. Immediately remove eggs and place into cold water, changing water several times until eggs have cooled.

- Combine all vinaigrette ingredients into a pint jar with a tight fitting lid, and shake to combine.

- When potatoes are done cooking, pour about half of the dressing over them while warm, to allow them to soak up the flavors.

- Cut the eggs into eighths, and add them to the potatoes.

- Add the bacon and toss the ingredients together.

- Taste, and add the remaining dressing as necessary. Serve with any remaining dressing on the side.

* If you are able, keep your farm fresh eggs refrigerated for one or two weeks before using them to make hard boiled eggs; they will peel much easier this way.

Farm Fresh Eggs

Eggs straight from the chicken coop and a basket of mustard greens and kale will soon be turned into a simple summer meal.

A Visit from the Vet

"Good veterinarians talk to animals.
Great veterinarians hear them talk back."

– Unknown

The mid-summer sky was overcast, and there was a slight breeze; it was a perfect afternoon for the vet to visit our herd for the first time and draw blood to test for the most common, and most serious goat diseases. The three breeders we had purchased our goats from all have good reputations in the goat community, so I hadn't felt the need to test for these diseases earlier. I felt fairly confident that the tests would be negative and would prove that we were the disease-free herd I expected us to be. Albeit, at two years in, and at the request of several potential customers, I had decided that testing our herd was the responsible thing to do.

The vet's clinic called and the receptionist on the end of the line told me that the vet would arrive in approximately fifteen minutes. I pulled my coveralls up over my shorts and tee-shirt. I hadn't gathered the eggs yet that day, so I went out to the chicken coop to collect the day's treasures. Just as I was walking back up to the house with a bucket of freshly laid eggs, the vet drove down the driveway and slowly rolled to a stop.

The vet extended his hand to shake mine and introduced himself by his first name; immediately, I felt at ease. He was soft-spoken and friendly. He asked about the animals I have and how long I've had them for. He mentioned that he also keeps several goats and a handful of hens, and that he was planning to test his goat herd this year as well but hadn't gotten around to it yet. As we entered the first paddock, he asked which goats were the most skittish—we would draw blood from these goats first.

I had brought several buckets of barley with me, and the goats clamored around me, eager to stick their noses into the buckets. I hung two buckets up for the does, and set a third up on a shelf to give to the bucks a little later. The two most skittish goats were the two doelings that we had kept on the farm from last year's kiddings: Lavender and Lilac. With the goats' muzzles deep in grain, I easily picked up the smaller of the two, Lavender. I brought her over to the milking stantion and sat down, holding her firmly in place on my lap.

The vet pulled out the first syringe and attached the tube to the end. As I gently tipped her head back, he felt for the jugular, and while he kept one finger on it, he inserted the needle with the other hand. She didn't make a peep. Dark viscous liquid pooled in the bottom of the tube and quickly filled it. And, just like that,

Our Nubian/Boer cross meat goats, enjoying their summer forage. They always come to the fence when they see us, hoping we've brought an extra snack.

he pulled out the needle, and I released my grip—she trotted off back to the barley bucket as if nothing had happened.

I followed her back into the barn and picked up Lilac next. She kicked a bit as I hoisted her into the air, but she didn't struggle. With a new syringe and a fresh tube, the vet quickly drew her blood as well. The rest of the goats were too big to carry, and so I put my arms under each of their front legs and walked them one by one out of the barn and up to the milking stantion, and then restrained them with a bucket of grain at their noses. None of them struggled much—and none of them yelped or squealed the way the vet had warned me that they might.

Once we had completed the does, we moved over to the bucks. I picked up the remaining bucket of barley and hopped over the fence. The vet followed, and soon our two herdsires and two wethers were happily munching away. I secured our smallest and most timid wether, Gingko; and as I knelt on the rough grass, the vet quickly took his blood. The larger and more friendly wether, Acorn, was next and he cooperated like a perfect gentleman.

Containing Rowan, our primary herdsire, was a little more difficult. He reared up on his muscular hind legs, struggling against my arms which were wrapped around his middle. I shifted positions and straddled his hind legs while leaning over his torso to hold his head up. Thankfully he submitted and allowed the vet and me to accomplish the task. As soon as I let go, he galloped to the top of the dirt hill in their paddock and reared up on his hind legs as if to remind us that he was still king of his castle.

Catching Hawthorn was a bit more complicated. He is by far the most skittish of all of our goats. I thought we had him cornered in his shelter, but he zig-zagged out right between us. After running after him for a couple of minutes, we realized that it was futile; we were going to have to strategize on how to capture him. We have a breeding pen set up in the corner of the bucks' paddock, and I swung the gate wide open in hopes that we could chase him into it. We chased him around the field one more time, and then just as I had hoped, right into the pen.

A Visit from the Vet

It took us just over half an hour to draw blood from nine goats. The vet's cargo pockets bulged with the tubes of goats' blood, and the wrappers and spent needles he used to draw the blood, as we walked back up from the field to his truck. He stopped to check out our chubby piglets as he walked past, and commented on what a good looking bunch they were. We joked about hoping we wouldn't see each other again until this time next year—implying that all would be well with the tests and our animals for the year to come. As he drove up the driveway, I sincerely hoped this would be the case.

Grape Jelly

"The juice of the grape is the liquid quintessence of concentrated sunbeams."

- Thomas Love Peacock

Unfortunately, grape vines take an agonizingly long time to grow before they produce any substantial amount of fruit. Our beautiful Cabernet Sauvignon vine is in its third year. It has crawled up a post that supports our deck and is beginning to weave its vines in and out of the posts along the railing. Last summer, we enjoyed one small cluster of grapes from this vine. This summer, I hope for more, but in reality it won't begin producing well until its fifth year. In the meantime, I am grateful for friends who are willing to share their abundance with us.

For the last two summers, friends have invited us over to harvest from their ancient Concord grape vine. The thick, gnarled vine has spent years—and likely decades—climbing up one of the beams and along the roof's edge of their covered deck. In the winter, it looks like nothing more than an overgrown, skeletal weed; but in the summer, when its broad green leaves and deep purple clusters of ripened fruit weigh down the vines, it is a breathtaking sight.

At the encouragement of our friends, we plucked off hundreds of clusters of dark, globe-shaped fruit, until the baskets and bowls we brought along with us were overflowing. The unusually warm, dry weather we have experienced this summer helped produce a multitude of incredibly large, sweet grapes. As tempted as I was to try my hand at homemade wine, I didn't want to waste even a drop of the precious juice on a rookie attempt that could so easily end in disaster. With a familiar taste of my childhood in the back of my mind, I knew that these grapes were destined to become grape jelly.

At home, a quick rinse to wash away any residual dust or insects was all that was needed before we began plucking each grape from the stem. My little sprouts helped with this part of the jelly-making process and turned it into a game; my oldest declared that the pot was the net and each grape a ball to try to get into the net. I pointed out the white yeasty bloom on the skin of the grapes and showed my youngest two sprouts how to carefully make a fingerprint in the yeast. My older two competed to see how far back they could stand and still get their grapes into the pot. Only a few landed on the floor, and I got my grapes into the pot in relatively short order.

The jelly-making process is not complicated, but it is time consuming, especially when making a large batch. Once my two largest stock pots were full, we turned on the heat to encourage the grapes to release their juices. The pots simmered on the stove for at least an hour, releasing huge plumes of steam into my already humid kitchen. I turned off the heat and let them cool down before straining the juice from the thick, pulpy skin and seeds.

I don't have the space in my kitchen or the patience for hang-

Grape Jelly

A pot full of grapes with the powdery bloom still on them.

ing jelly bags, so I used the next best thing to strain out the pulp—a stainless steel colander. From the pot, through the colander, and into an extra-large glass measuring cup, I filled as many half-gallon and quart jars as it took, until every last drop of juice had been squished out of the pulpy leftovers. It's a good thing we have a second fridge that I use to store our eggs; by that evening, half of the fridge was filled with freshly pressed grape juice. I planned to leave them overnight to allow the sediment to settle on the bottom of the jars; I would tackle the jelly-making process in the coolest hours of the early morning.

My youngest sprout woke up before me the following morning, and I awoke to a large crash and a wailing three year old. I ran downstairs to see what the commotion was all about, only to see my startled little girl standing in a massive puddle of grape juice, with broken glass scattered all around her feet, and huge tears rolling down her cheeks. I hoisted her up out of the puddle and carried her to the bathroom sink to wash off her feet, and to be sure she wasn't hurt. By the time her feet were clean, she had calmed down enough for me to ask what had happened. I wanted to taste your jelly, she replied.

I used towels to barricade the juice from crawling further along the floor, and then proceeded to clean up the pieces of broken mason jars—one half gallon and one quart jar. It took everything in me to hold back the sob I felt rising up from my gut. I felt incredibly frustrated by the mess, but also by the wasted juice. It was only my second attempt at grape jelly, and it was not off to a good start.

Once the glass had been cleaned up, my preschooler and I began to sop up the rest of the three quarts of juice that stretched out across my stone floor. I was thankful that at least it had happened in our laundry room and not in the kitchen; I was sure there was no way the deep purple stains would be coming up. As I wiped the juice away, I could see that the acidity from the juice was lifting off the seal on the floor, and soaking right into the stone tiles.

In a panic, I called a friend who always seems to have the right cleaning tip for the job. She suggested washing the area with diluted bleach. Seeing as the tiles were already ruined, I figured that bleach couldn't damage them any further and gave it a try. I was amazed to see that the stains lifted right out of the porous tiles. By the time Mr. Green Thumb got home that evening, our laundry room floor looked as good as new; all that remained was the story about crying over spilled grape juice and my sore biceps from scrubbing the floor clean with bleach.

I couldn't stand the thought of looking at anymore grape juice

A cluster of grapes ripening on the vine on our homestead.

for the rest of the day, and so it wasn't until the following day that I made the grape jelly and my youngest was able to finally taste it. I hoisted her up onto my hip and reached for a spoon to scrape out the last bit of jelly clinging to the sides of the giant stock pot cooling on my stovetop. A big grin crept onto her face as she licked the spoon clean. "More?" she asked, innocently holding up the spoon. I had to laugh as I gave her another taste of the jiggly purple jelly.

Oh little one, you have no idea the stress I went through to make this jelly, I thought to myself with a smile. Her cherub face beamed as she licked the spoon clean once again.

Grow. Cook. Eat. Share.

Grape Jelly

Ingredients:

4 cups Concord grape juice (homemade or store bought)
*(see tip below)
2 cups sugar
1 package pectin of choice
(I prefer Pamona's Universal Pectin)

Instructions:

- As the jam and jelly making process goes quite quickly once started, measuring out all of the ingredients and having all the tools necessary beforehand is advised.

- Wash and dry four half pint or two pint jars and metal bands; (make sure you have new lids to fit the jars).

- Fill a large stock pot, or canning basin about half full and bring to a boil; you want the water to cover the jars while they are being canned.

- Read over pectin instructions for making Concord grape jelly, and follow accordingly.

- If using Pamona's pectin, prepare the calcium water in a jar, close, and shake well.

- Measure out 2 cups sugar and pectin powder according to the recipe. Thoroughly mix the pectin powder into the sugar.

- Measure juice and calcium water into a large sized pot and stir well; be sure the pot holds, at least double the volume as the juice mixture, as it can increase in volume once the sugar is added and then heated, and you don't want the hot liquid overflowing the pot!

- Bring juice to a boil. Once boiling, add the sugar and pectin mixture.

- Continuously stir for two minutes to dissolve the sugar and pectin, while juice is still boiling.

- Take pan off the heat.

- Immediately fill clean jars to ¼" of the top; wipe off the tops of the jars, and place the lids and the metal bands on. Screw on the bands on firmly but not tightly.

- Place jars into the boiling water, and boil for ten minutes; (For every 1,000 feet above sea level, add another minute.)

- Remove from water and check seals; after an hour, all the jars should have sealed, and the lids will be sunken in.

- Any jars that do not seal need to be placed in the refrigerator, and used within three weeks.

- Allow to stand for 24 hours; disturbing the jars early may cause the seal to break.

- The sealed jars will last at least one year in your pantry.

* This recipe can be doubled, tripled, or even quadrupled; just be sure to multiply all of the ingredients, follow the instructions in the pectin package, and have enough jars and supplies on hand, as well as enough time to complete the canning process.

How To Do It All

"I am always doing that which I cannot do, in order that I may learn how to do it."

- Pablo Picasso

At this point, you may be asking yourself the very same question my friends ask me every time we are together—how do you do it all? This is a valid question, because from an outsider's perspective, homesteading appears to be a physically demanding, 24/7 job, with little to no time off. However, I would ask the same question to the women and men who have full-time jobs and tiring commutes, only to come home and cart their children to and from multiple sports practices and games, as well as trying to fit in grocery shopping and socializing.

If you are still asking, but how, I'll let you in on two secrets: the first one being that most of the time, I love what I do. Not only is the work I do my job, it is my hobby as well, and I genuinely enjoy it. I love watching the chickens scratch and peck around their run, clucking and squawking at each other. I love hearing the punctuated egg song each hen sings after she has laid an egg. It's her proud moment in the spotlight as she lets the world know she's done her job. I love seeing Mr. Peabody, our peacock, strut around our property, fanning out his tail feathers for the ladies and any humans that may be nearby.

I love watching my youngest, CJ, head down to the goat pasture after school most days. Never mind the fence—she shimmies over it; and before her feet hit the ground, the herd is hovering around her looking for ear scratches and a rousing game of tag. I love that I get to witness the miracle of life year round—watching day old goat kids suckling at their mama's teats; watching two month old piglets frolicking in the field; watching week old chicks and ducklings toddling after their mamas through the long grass.

I love kneeling in the garden with the sun warming my back and my hands in the dirt. I love that even before I can be in the garden, I can cozy up in front of the fireplace with a handful of seed catalogues and dream of the day when I can once again dig in the dirt. I love going to the garden each afternoon during the summer months and harvesting the vegetables needed for that night's supper; I love eating a sun-kissed strawberry or a vine ripened tomato straight from the plant as I harvest the abundance our garden produces.

Most people have a job, a hobby or two, an exercise routine, and a list of daily and weekly chores. Almost everyone regularly has to go to the grocery store for basics like meat, dairy, vegetables and pantry staples. We have the same demands here on our homestead, but the difference is that we don't have to leave our property to do a lot of these things. It's true that Mr. Green Thumb still goes to work off our property every day, and the kids are in school—we're so fortunate that there is a great school just up the road from us—but our hobbies, our exercise, our grocery shopping, and our entertainment are mainly accomplished right where we live.

My routine has definitely become more relaxed than when I

The unpleasant but necessary job of cleaning out the shavings from the chicken coop. The shavings will age in our compost pile before being added to the garden beds.

lived a suburban life. I'm not afraid to admit that I don't always shower every day, and only wash my hair once or twice a week; it's better for my hair and for the environment, right? No blonde highlights for me anymore, or nail polish, or chic outfits on a daily basis. Often I do my chores in my pajamas, only to turn around and run my kiddos to school in said pj's. It's not that I've stopped caring—I most certainly dress nicely on date nights with Mr. Green Thumb, or when I meet my friends for coffee, or when we go to church—it just doesn't feel as important anymore because

I've learned by living on our homestead that I am so much more than simply how I appear on the outside.

I am proud of my body for all of the things it does for me. I love that I can carry two five-gallon pails of feed or water without breaking a sweat. I get giddy loading up the wheelbarrow with aged manure to top dress my garden beds, or filling the hayloft with seventy-five pound bales. I can get out my frustrations by attacking the blackberry bushes with my machete, and I impress even myself on the days I clean out the deep bedding in the chicken coop with nothing but a pitchfork, a wheelbarrow and sweat-drenched will power. I may not jog on the treadmill or lift weights at the gym, but I am certainly exercising as I go about my chores.

The second secret I'm about to let you in on is more difficult for me to admit—my house is almost never clean. Don't get me wrong, it's not as if we live in a barn—but let's just say that when the sprouts see Mr. Green Thumb and me frantically cleaning, and prodding them to do the same, they ask who's coming over. Often I've cleaned one room, only to turn around to a new mess in the next room over. I know this has to do with having four young sprouts, and a lot of dirt around the homestead, and I'm hopeful that one day the house will stay a little cleaner; but while I do my best, I make no promises—house cleaning simply does not bring me joy. When I saw a sign reading Gardening Forever, House Cleaning Whenever on my mother-in-law's back porch, I knew I had married into the right family.

I know that there are homesteaders who keep their houses spotless. Some people find great satisfaction in seeing all of their clean laundry folded, their bathrooms sparkling, and their kitchen sinks empty. While I also feel happy knowing that my dishes are

The first harvest of apples from one of our young apple trees.

clean and my sprouts have clean clothes to wear, I recognize my limitations. I only have so many hours in a day to cook, clean, write, finish the homestead chores, and still spend a little quality time with my family. Even with the help of Mr. Green Thumb and the sprouts, we still aren't able to accomplish everything on this list every day, and something has to give. I have prioritized my family and our homestead over dust-free bookshelves and crumb-free floors, and I trust that when we have guests over, they would rather enjoy a home cooked meal around the table than criticize a sink full of dirty dishes used to make the meal.

As much as I get done in a day, I have just as much on my list that doesn't get accomplished. Over time, I have learned to make peace with the idea that I cannot complete everything on my homestead list. I know what needs to get done, and I prioritize those things. Animals and sprouts get fed; clothes and dishes get washed; the garden gets watered and weeded when it absolutely needs it, and the eggs are gathered. Some weeks we bake our bread, and some weeks we buy a loaf or two. Some winters our pantry is packed full, and others, we supplement our canning with a few purchased jars here and there. Taking the pressure off of myself in this way has actually enabled me to get more done, rather than worry my time away, fretting that my list won't be completed.

I am also grateful that Mr. Green Thumb respects my limitations, while recognizing his own. With him working away from the homestead on weekdays, he cannot accomplish everything he would like either. And so we work hard, and make some sacrifices, and find ways to create quality time with the sprouts while still working, but at the end of the day, we rest, knowing we have done the best we could. There will always be tomorrow to get more done.

A jar of beets in brine; these will soon become fermented beets, and the liquid, beet kvass.

The Farm Cats

"If the pull of the outside world is strong, there is also a pull towards the human. The cat may disappear on its own errands, but sooner or later, it returns once again for a little while, to greet us with its own type of love."

-Loyd Alexander

Did you know that taco cat spelled backwards is taco cat? I didn't either until my two oldest sprouts pointed it out one day on the way home from school. Shortly after this illuminating conversation, we adopted an aloof, orange mouser from a farm family that was moving to the 'burbs; we named him Taco, obviously. Since Taco, the cat population on our homestead has exploded, turning into a revolving door of kittens, tom cats, queens, pregnant mamas, and more kittens. The four cats that moved to our homestead, shortly after Taco did, were a brother and sister named Cosmo and Sugar Cube, and two sisters named Mocha and Latte.

After months of frolicking outside, proudly dropping mice off at our doorstep, and being all around adorable, the three females went into heat. From the comfort of our beds at all hours of the night, all six of us could hear our queens yowling and screeching in response to the neighborhood tom cats. For weeks this continued, and on several nights, our sprouts burst into our room, scared of the noises they heard from beyond the darkness. We informed them that there was nothing to fear—those sounds were ours and the neighborhood cats getting married and making babies.

While Latte and Mocha didn't want anything to do with us during their time in heat, our petite, jet-black beauty, Sugar Cube, tried to get into the house continually. I finally brought her in, placed her into a pet carrier, and called up my friend to ask if her orange tom cat would like to have a date. She agreed and we dropped her off later that afternoon. My friend had already placed her handsome ginger boy into an oversized pet crate, and Sugar Cube was me-owing as we carried her closer. I didn't have time to stay and see if they were interested in each other, but my friend offered to send updates on whether or not they got along.

That evening, she texted to say that Sugar Cube had been continuously swatting at her boy, and her usually strong, bold male was cowering in the opposite corner of the crate. We agreed to leave them overnight to see if they had changed their minds by morning. However, another tom cat had come to check out what all the commotion was about. The next morning, my friend found him and Sugar Cube resting against each other on either side of

I'm holding our sweet boy Cosmo.

the crate, purring contentedly. It was a perfect match. My friend switched out one tom for the other, and Sugar Cube was thrilled with her new mate. They cuddled together for the rest of the weekend. We were pretty sure that all three queens were now pregnant.

Mocha, Sugar Cube, and Latte all gave birth within a week of each other, and we went from three expectant mamas to seventeen cats and kittens in seven days. Shortly after our last cat had given birth to her litter, Mr. Green Thumb and I attended a wedding. I couldn't remember the last time I had put on nylons, let alone in the middle of summer. As we were saying goodbye to the sprouts, dressed to the nines and already sweating in the midday heat, our youngest, who was five at the time, upon hugging me, patted my tummy and suggested that there might be a kitten in there. I couldn't help but laugh and let her know that there were certainly no kittens in there, or anything else for that matter. However, I suppose I could at least be happy that she thought it was only one kitten and not a whole litter.

Going into winter, we had fourteen cats living on our homestead; only four cats survived the great coyote massacre our farm experienced this past year: OJ, our gentle, orange, male tabby; Divot, a grey-blue spitfire who has no interest in human contact; M&M, short for Mini Mocha, as she is the spitting image of her brown tabby mama; and Sundae, a brown and grey striped cat, who is even more timid than her sister, M&M. In early spring, as I was feeding the cats in the garage, I noticed that the three girls who were born on our homestead just last spring, had protruding bellies; I told the kids that afternoon that we could expect kittens in about a month.

For several weeks, we had been anticipating the arrival of kittens. Would the mamas give birth in the garage or in the house like last year? the sprouts pondered. As I was busy working outside one day, Divot, the roundest of our three mamas-to-be, came up to me and started rubbing against my legs. It had been months since I had gotten close enough to pet her; she always ran from anyone that tried to get too close. I had a hunch that her uncharacteristic behavior was an indication that her time was getting close. I carried her inside, and the girls got busy setting up a maternity ward in our dining room so that Divot would be comfortable while she waited for the big day.

Taco cat loved lounging on top of the piano in the warm, southern facing office.

Mr. Green Thumb and I were en route to pick up one of our sprouts from the school play he was a part of, when we got a panicked call from our oldest—"I think Divot's water broke! When are you coming home?"

We arrived home as quickly as possible, and learned that her water had indeed broken—all over our living room couch. Armed with rags and a bottle of my homemade vinegar cleaning solution, our oldest got to work scrubbing while I went to inspect our mama-to-be. She had found a safe spot on the bottom corner of our craft shelf on the opposite side of the room from the makeshift maternity pen. The sprouts led us to her hiding spot, and proudly pointed out that they had tried to tuck a towel under her. I glanced in to see how she was progressing, and saw a tiny, black and orange fur ball, still wet, with gooey placental liquid dripping all over the pile of construction paper that mama and baby were lying on. The towel had missed the mark.

My two youngest sprouts were hovering behind me, and I instructed them to wash their hands so they could help me as self appointed cat-midwives. My youngest, CJ, came back first, wiping her cold, damp hands on my bare arm to prove to me she had done as I asked. Gently, I placed the precious baby, smaller than her six year old palms, into her eager hands. She cradled the kitten up against her bosom, and instinctually bounced and rocked her tiny charge.

I could only imagine what Divot was feeling as I gingerly lifted her so that I could slip the towel under her bulging frame; I would have been unimpressed with being moved if I were in her condition, and I'm sure she felt the same, although she didn't resist. Once she was settled again, we placed her firstborn next to her, and it instinctively began searching for a teat. CJ, in typical dramatic fashion, turned to me and told me that she just couldn't wash her hands again, now that she had held the baby kitten.

I was still sitting on the floor next to Divot and was privileged enough to witness her push out her second kitten—another black and orange tabby. The girls both clamored around me trying to catch a glimpse as well, but it happened so quickly that by the time I had shifted out of the way so that they could get a better view, it was over. Divot began furiously licking kitten number two and gnawing on the umbilical cord to sever it. The girls felt squeamish as they watched Divot eat the entire placenta; and they were understandably horrified to think about the fact that most animals do this.

And then, the questions came: Why do animals eat their placentas? Why do some animals eat their babies? Do newborn kittens pee and poo? Why is Divot's tail bloody? Why did you just say v****a? I answered as best as I could, but in that moment, they weren't interested in my replies; they were attempting to make sense of what was happening right before them, and they needed someone to listen to their verbal processing as they did so.

Aussie, cuddling one of our kittens, Tiger.

I placed the first kitten, now mostly dry, into Rae-Rae's hands as Divot continued to lick clean her second babe. My sprout swooned as she caressed the fur on its head and down its spine, its toothpick-sized tail flailing, its mouth rooting around her fingers with eyes squeezed shut. It pained her to put the little one back down, but she knew it needed its mama's milk. It was late, and my girlies were tired. As I lay down together with each of them to say goodnight, they rambled on about what they had just seen, their yawning and restlessness eventually giving way to sleep. It had been another long day on the homestead but one I imagine my sprouts won't tire of remembering.

Summer Cordial

"When Anne came back from the kitchen Diana was drinking her second glassful of cordial; and, being entreated thereto by Anne, she offered no particular objection to the drinking of a third. The tumblerfuls were generous ones and the raspberry cordial was certainly very nice."

- L. M. Montgomery

    I am continuously on the lookout for new—or old, really—ways to preserve food. When I stumbled across a new-to-me method to preserve the sweet taste of summer's fruits and berries in strong, dark rum, I simply had to test it out.

    The tradition dates back several hundred years and is said to have begun when sailors along the trade routes began bringing rum to Europe from the Caribbean islands; the sea dwellers looked for ways to preserve the aromatic tropical fruits of pineapple and mango to share with those at home, and added the fruit to the rum for safekeeping during their time at sea.

The recipe I used to preserve my fruit in rum is called rumtopf, German for rum pot, and it can be adapted well to whichever summer fruits thrive in your particular growing region. Making rumtopf is as simple as dicing up a variety of fruits into a glass or ceramic vessel and covering the fruit with ninety-proof rum and a generous helping of sugar. I first began the process of making rumtopf last June as I layered homegrown, sliced strawberries and plump, whole blueberries into a two quart mason jar. I then sprinkled in a couple of tablespoons of raw cane sugar and splashed the viscous, nose-tingling dark rum into the jar—just enough to cover the fruit. I tried to remember to give the jar a swirl once or twice weekly as my memory permitted. As the weeks passed, the fruit grew darker and became saturated with rum, and was then floating at the top of the liquid.

Next, I added the diced peaches and whole cherries and raspberries, followed by blackberries, plums, apples, and finally pears. I pressed down the diced fruit with a spatula and spooned in a little more cane sugar. The syrupy liquid rose in the jar as I stuffed in the last of the fruit just below the rim of the jar; my bottle of rum was almost empty.

A scene from Anne of Green Gables played out in my mind as I labeled the jar with a pen made for writing on glass: rumtopf, fruit cordial, alcohol. I couldn't imagine the horror that beloved, gentle Anne had faced when she realized the mix-up she had made and had indeed gotten her dear friend Diana drunk on raspberry cordial. I did not intend for any of my sprouts to make this same mistake, and so I added the words, DO NOT DRINK for good measure. With one last shake of the jar, I placed it into the back of the pantry cupboard where the fruit and rum would

Our black raspberries are beginning to ripen. They are firmer than other raspberry varieties and are an excellent addition to rumtopf, or for eating straight off the bush.

mingle and marinade for several months. I looked forward to the Christmas season, when I would surprise my guests with this homemade, adult-only fruit compote atop pound cake, accompanied by a sweet, after-supper cordial.

The waiting was torture. I knew that gazing at the jar as I turned it over in my hands would not speed the process along, but I couldn't help myself, and I periodically pulled the jar from the shelf and held it up to the light, giving it a slight swirl as I did so.

The fruit was as dark as the rum, and seeds from the berries had settled at the bottom of the jar.

Some of the softer fruits were not holding up as well as the apples, cherries and plums but were beginning to break apart; and yet, the superior preservation ability of the alcohol prevented any unhealthy bacteria or mold from growing inside the jar. The juice from the fruit, mixed with the sugar and the rum, was creating a completely unique beverage, specific to the fruits I had grown on our homestead. This delicacy was as close as I had come to tasting my own terroir in a bottle, and the anticipation was intoxicating.

When December rolled around, I couldn't wait any longer. One weeknight with nothing special going on, except that the sprouts were already down for the night, Mr. Green Thumb and I taste-tested my long awaited experiment. The dark amber cordial was syrupy-sweet, much like any dessert wine, and required just the smallest sips out of the mix-matched pair of antique, crystal goblets we held. I licked my lips, wiping away the sticky, sweet liquid after each taste. The flavor was fruity but subdued and elegant. It tickled my tastebuds and warmed my throat as it went down. It was the perfect night cap for a cold, blustery evening.

I felt my stomach flip-flop at the thought of consuming the fruit next—fruit that had sat on a slightly cooler than room-temperature shelf for the last four months—never mind that it had bathed in alcohol and sugar syrup. I was no stranger to consuming fermented foods that had happily percolated on my counter for a month or two with nothing but lactic acid standing between me and botulism, but this form of preservation was new to me, and I felt nervous as I took my first taste.

I had no reason to be worried though; it tasted wonderful.

Our strawberry plants thrived, and their fruit grew bigger than we had ever experienced, once we added manure from Aussie's 4H rabbit.

The fruit was soft as though cooked but still maintained its distinct characteristics; the diced apples and cherries still tasted like apples and cherries, although sweeter and punchier than before. The stone fruits were delicately soft, but each was recognizable. The raspberries, strawberry slices, blackberries and diced pear fared the worst, but still gave body and flavor to the syrup and compote, which we enjoyed that night, ladled over a humble scoop of vanilla ice cream. I'd like to say that our friends enjoyed it as well, but it disappeared before they had the chance.

Summer Cordial (Rumtopf)

Ingredients:

Add to a two-quart jar as fruit is in season (at perfect ripeness, and free of mold and bruising):

¾ cup blueberries

¾ cup hulled, halved strawberries

1 peeled, diced peach

1-2 diced Italian prune plums or other plums

1 cup pitted cherries

¾ cup raspberries

¾ cup blackberries

½-1 peeled, diced pear

½-1 peeled, diced apple

Feel free to substitute any fruit listed above for other fruit you may have available, or try making a tropical version using pineapples, mangos, dragon fruit, passionfruit, etc. (Just don't use creamy fruits such as banana or papaya.)

Approximately 4 cups sugar

45% or 90 proof golden/ dark rum (not spiced)
(Since I haven't been able to find a 45% dark rum, I use a combination of two types of dark rum with different percentages to equal 45%.)

Summer Cordial

A handful of freshly picked blueberries, in front of the purple delphinium flowers that border our garden.

Instructions:

• As you add fruits to the jar, add ½-1 cup sugar to the jar as well, and then pour enough dark rum into the jar to cover the fruit. Leave it on the counter and add fruit, sugar and rum as each fruit is available.

• Once the jar is full of fruit, sugar, and rum, place it in a dark, cool location and let it sit for approximately two months.

• Then, open, serve, and enjoy!

The Straw that Broke the Goat's Back

"Animals have a much better attitude to life and death than we do. They know when their time has come, We are the ones that suffer when they pass, but it's a healing kind of grief that enables us to deal with other griefs that are not so easy to grab hold of."

- Emmylou Harris

"I am so done," I exclaimed, as I collapsed onto Mr. Green Thumb's chest, the moment he walked through the door. Tears began to trickle from the corners of my eyes for a second time that day. That afternoon, I had received a phone call from our vet to let me know that two of our does, a mother daughter pair, had tested positive for a goat disease called CAE. I was shocked. Our animals had all come from reputable breeders who take care to test their animals and keep them healthy. As I ended the call from the vet, I felt like the carpet had been pulled out from under me.

CAE is short for Caprine Arthritis Encephalitis; while positive goats don't necessarily display signs of the disease, the onset of symptoms wreaks havoc on a goat and will inevitably lead to a painful, premature death. The worst of this disease is that there is no vaccine, no cure, and no humane way to deal with the disease other than culling the affected animals. On a positive note, the disease cannot live long outside of a host, and once the affected animals are removed, it is a short incubation period before a retest will determine whether any other animals have been infected.

While many farmers keep CAE positive animals, and only cull at the onset of the disease, I felt that would be an unnecessary risk to take, considering that our acreage is small, and the risk for cross contamination between feed and water buckets, and even simply cuddling the infected does, would present; culling would be the best preventative measure to ensure a clean herd, free from disease.

I had made the hard decision to cull, even before we had purchased our first goats, if disease or injury were to happen. I wasn't expecting it to happen within the first two years of keeping goats though, and it caught me unprepared. It took almost a week after the pair had been quarantined before I was able to ask the vet to come and put them down.

When the vet arrived, I shared my thoughts about why I wanted to preventatively cull as he gathered his materials out of the back of his truck. His insights into how quickly and painfully CAE can manifest itself, and his experience witnessing the disease, was confirmation that this was the most humane decision. As heartbroken as I felt to lose two beautiful animals that I had spent over a year nurturing and tending to, it would be harder for me to see

The Straw that Broke the Goat's Back

Our dam Violet, licking off her doeling Lavender, just hours old.

that I had caused them unnecessary pain.

The walk down to the temporary shelter we were using to keep them quarantined was slow and somber. The silence hung between us, and we each measured the weight of the task at hand. We quietly entered their paddock, and I knelt down to give them each a cuddle as they nuzzled up to me—I was painfully aware that I would need to wash these clothes before going in to check on my other goats. The vet used a calm soothing voice, for me just as much as for the animals to be sure. As he pulled out the first syringe, he explained that he would first sedate them.

I held Violet close—one arm around her middle and the other around her head—as he plunged the needle into her neck. She fought a little, but once I let go, she continued to stay by my side as I caressed her ears. Lavender fought harder. She didn't want to be contained and squealed as our vet gently pressed the needle into her flesh. She was much calmer when we had drawn blood several weeks earlier—maybe she sensed our inner distress. "It will take several minutes," the vet said. We watched them stumble around, before collapsing, with our help, onto the hard ground.

He listened to their breathing to be sure they were sound asleep before administering the lethal cocktail in their neck. First Violet, and then Lavender. They didn't flinch. I stroked Violet's head. I didn't want to say goodbye like this, I whispered. I thought I would feel emotional; I thought I would shed tears. But I felt nothing, and I didn't like that. It shouldn't have been so easy to end two lives. I should have felt more, I chastised myself. The sky was getting darker, and then rain began to pummel the tarp above our heads.

I was more disturbed by bagging them up. They each fit so perfectly into the black garbage bags I had brought with me. These precious lives bagged up as nothing more than biological waste. Their carcasses were warm and floppy; trying to wiggle the bag up around each of them took effort, and I struggled against the bag until the vet came over and finished the job. We double bagged them, and then locked the gate on the way out of the temporary paddock, to prevent any wildlife from getting to them before morning. Mr. Green Thumb would deliver them to the deadstock drop-off the following day.

I didn't have the heart to tell our sprouts until the next morning when Mr. Green Thumb drove his truck down to the goat paddock. Our youngest insisted that she come out to help her daddy with whatever he was doing on the farm. When I broke the news to her, and suggested she might not want to be there, she immediately began to wail, which had her three older siblings running to her side to see what was wrong.

I told the other three that vet had put down Violet and her daughter Lavender the day before, and then there were tears all around. This bought my husband time to load up their carcasses and drive away, uninterrupted. My four sprouts and I hugged, and we talked about what good goats they had been, and how this seemed unfair, because it wasn't their fault or ours that they had gotten this disease. They asked how the goats had died, and I explained that the vet had been very gentle, and had put them to sleep first so that they hadn't felt a thing.

As I sit back and evaluate all that has happened in the last four years, I realize that we've ticked off all the boxes of our homestead dreams. Large garden, orchard, berry patch, goats, chickens, pigs —even a peacock. Looking at hens meandering around the fruit trees brings a smile to my face. When I see the peacock in full display, I marvel at his beauty. When I watch the baby goats bounce around the field next to their mamas, I feel bliss. Yet in these four years, I have not stopped to count the cost—both positive and negative—of what this homestead life has done to my emotions, my home, and my family.

As Mr. Green Thumb has become more consumed in his business off the farm, I have taken almost all of the farm tasks onto my shoulders. Not just the daily chores of feeding and watering, but also the larger activities of cleaning out the shelters, ordering feed, pruning our fruit trees and berry bushes, trimming hooves, planting and weeding; and the list goes on. Day after day the outside chores have been prioritized over the inside chores, until our whole family has become blinded to living among piles of clutter and baskets of unfolded laundry. The weight of the mental load that these tasks have added has pushed me down into a dark place—a place that would rival a pig's mud pit. Not just physically, but mentally, I am exhausted.

This mental burden I am carrying has insisted that I pause and reevaluate what I am doing and the direction I am headed—not only as a small scale farmer, but as a wife, a mother, a person. The joy that I felt as we continued to check off each box on our list has been diminished by the painful presence of a mountain of tasks at hand. We have grown, added, and increased, to the point where I am now teetering on the top of a very unstable tower; a tower that is both my doing and my undoing, and there is only one way off—down.

The mental load of the goats, chickens, rabbits, pigs, and garden weighs on me daily, and I am struggling to remember why this lifestyle made sense to me just last year. And so, I am in the process of determining what down must look like for me. I certainly cannot maintain the downtrodden state I find myself in. Deep in my gut, I know what I need to do, but I am having a hard time admitting it, even to myself. Downsize—this is what I must do.

Violet gazing at me with her gentle blue eyes.

EAT.
Autumn

Autumn Days

"Autumn... the year's last, loveliest smile."

- William Cullen Bryant

Autumn is a season of opposites; cold northeastern winds juxtaposed with cozy sweaters, scarves, and hats; empty fields and animal shelters, contrasted with overflowing larders and freezers; ruby, amber, and topaz leaves clinging to their branches against the backdrop of grey, overcast skies; the snow capped North Shore Mountains and the flickering flames in the fireplace; the end of a dizzying hustle, the beginning of a time to rest; the pantry lined with mason jars, and yet, my need to not see another mason jar for weeks.

Autumn has furiously rushed in to replace the last days of summer; the crisp evenings send a chill down my spine. The tender leaves of the basil, lettuce, cilantro, and tomato plants have not been able to withstand these temperatures and have shriveled and browned before crumpling to the ground. The tender plants are gone but the hardy mustards, kales, cabbages, and brussel sprouts are thriving. Almost daily we are eating brassicas from the garden, incorporating them into stews, soups, and pans of roasted vegetables. It's the start to the season of warming, hearty comfort food, and these greens are making their way into many recipes.

Most days the clouds hang low and the rain is ever looming when it is not falling from the sky. In this corner of the world, most autumn days bring rain, whether it is a light drizzle, a rain shower, or a downpour. There is no longer a need to fill the animal waterers, as the rain keeps their buckets full. However, it is necessary to break apart the icy layer on the surface of the water on the nights that drop below freezing, and to prevent the food troughs from filling up with rainwater, turning the animals' pellets into a sludgy soup.

The kids are settled into their school routines, and I miss having the help and the company of my four sprouts as I walk from the garden, to the chicken coop, and to the barn, caring for the plants and animals. They come home from school tired and in need of some quiet time to decompress from the noise, the busyness, and both the physical and emotional energy it takes to navigate their teachers' and peers' requirements. They each take an hour or so to themselves before I remind them of their daily chores, both inside and out.

On the few dry days we have in mid-October, I make time to plant my seed garlic, encouraging each one to send out its earliest roots before the ground freezes over for the winter. The sprouts and I spent time this past summer separating the bulbs of garlic after they had cured, so they would be ready to go into the ground

A beautiful head of cabbage waits in our garden, blanketed by an early frost.

once the narrow window of planting arrived. Row after row of plump Russian red cloves have been buried into well aerated soil, teaming with worms and all sorts of microscopic organisms. I chose to plant bare-handed as I often do; I love the feel of the soil between my fingers, the aroma of the living earth wafting up as I dig down to make a hole for each clove.

The hens are molting again—didn't they just do this? I can't believe it's already been a year since they went through their last molt. Egg production is down from two dozen this summer to four or five a day. I only have eggs for one customer now, and I'll make sure it's one customer in particular who gets them. She has been a loyal customer of ours for years and relies on our pastured, soy and corn free eggs as a part of her diet to help manage her autoimmune disease; it's important that she gets the eggs she needs, as much as we can provide them.

Many weeks this autumn, I've only kept a couple of eggs as needed to make a batch of waffles or muffins. We haven't eaten scrambled or hard boiled eggs in over a month. This spring, in the midst of our abundant egg season, I had cracked several dozen eggs into a bowl, whisked them up, and then poured them into ice cube trays to freeze. However, six months had taken their toll, and the eggs, even cooked, tasted awfully freezer burnt. In years past, I have resisted buying eggs, but this year I caved. I ended up buying a couple of dozen organic eggs from our big box store while I waited for our hens to finish growing in their new feathers.

We took our two lambs to slaughter just last week. Saying goodbye wasn't easy for the sprouts as it was our first time raising lambs, and their fluffy fleece and sweet faces were still so young and adorable. I named them Rack and Of to remind our sprouts that these lambs' sole purpose was to eventually feed our family. Regardless, it was a challenge to let them go. In hindsight, I don't think the kiddos had ever eaten rack of lamb before, so the names may not have had the impact I had hoped for. I, on the other hand, was very excited to get my hands on all of the meat we would receive back. I had visions of lamb chops dancing in my head as I eagerly waited to hear back from the butcher about when our cuts would be ready.

Mr. Green thumb was considerate of the sprouts' struggle, and came home early from work one day so that we would be able to

"Rack" and "Of", the lambs we are raising for meat.

drop off the lambs before the kiddos returned home from school. On the way through town, we drove through a Starbucks drive-through to pick up a coffee before hitting the highway to get to the slaughterhouse. I laughed as I related to Mr. Green Thumb that I had also driven through a Starbucks drive-through with the lambs in the back of the truck, the day I had picked them up. Clearly, it's one of my guilty pleasures. It also made for good conversation with each of the baristas who greeted us at the window.

    I couldn't believe my luck several days later, when I happened to be at the butcher shop the day that the truck showed up with our lambs from the slaughterhouse. I asked the butcher if he would mind me coming into their walk-in cooler to see the lambs before they processed them—I was so curious to see what they looked like, and how they differed from our pig carcasses we had seen hanging a few months back. The butcher cheerfully agreed, and I followed him to the back of the building like an excited puppy; I could hardly contain my exuberance.

I saw pig, lamb, and beef carcasses hanging in rows as we stepped into the cooler. The butcher informed me that along with the meat they sold at their shop, they were busy processing 4-H animals as well. I admired the ruby red flesh of the hanging carcasses as we walked down the aisle. I knew how much work had gone into raising each animal; how precious each life was; how appreciative each family would be as they enjoyed the steaks, loins, and chops of these well raised animals. And then we were standing in front of our own lambs—the ones we had nurtured, and fed, and given ear scratches to. I closed my eyes and inhaled deeply in gratitude for the sacrifice of these animals.

"They have a really deep maroon color to the flesh," said the butcher, and I snapped to the present. "You can tell they were really healthy, happy animals," he continued.

I beamed. Yes, exactly! I wanted to shout. All of our time spent caring for these creatures, would pay off in feeding our family with their deeply flavorful, nourishing meat. I was thrilled with the compliments the butcher had paid to our lambs. Before pulling out of the parking lot, I sent a quick text to Mr. Green Thumb letting him know what the butcher had said. I would be even more impatient now, waiting until the meat would be ready to pick up.

The coyotes have been busy these days, since their easy prey of wild rabbits and other tiny mammals has become scarce. Up

A row of garlic drying out after harvest, so that it can be stored through the winter.

until now, they hadn't accessed our property—we only ever saw then pacing around the outside of our fence line on the neighboring properties. This autumn however, we have seen them boldly strolling down our driveway, keeping an eye out for a foraging hen, a barn cat, or anything else they can sink their teeth into. Within several months, we have faced an avalanche of losses to these pesky predators—eight of them being animals our kids loved and considered pets.

One of our mama cats and then four kittens disappeared one by one; then three goat kids—two in one night, and the third about a week later. Forty hens in total went missing, although I only realized this once our birds started laying again, and we were still only getting about a dozen eggs a day. Our oldest sprout is the one feeding and tending to the birds these days, and I hadn't thought to have him take a headcount of our hens. Most recently, our handsome peacock Mr. Peabody was snatched. We noticed two piles of his feathers in the neighbor's field along the road as we drove past. I really miss seeing Mr. Peabody strutting around the property.

Our sprouts are handling the loss of their pets and our farm animals better this year than they have in the past. In part, because they are getting older, but also because they have experienced so many animals disappearing, dying, or being sent to the butcher over the five years we have lived here. I recently asked my oldest, AJ, what he thought the benefits were of living on a homestead. He was silent for about a minute and I could see the thoughts churning in his mind, maybe even feeling hesitant to speak what he was thinking. And then, my thoughtful son solemnly told me that our homestead has prepared him, and all the sprouts, to better handle death. I have had several follow-up conversations to be sure that there is no resentment or trauma in regards to losing our livestock and pets, and it appears that they are doing well in this regard; I have since made a point to learn from my sprouts on this matter.

A wheelbarrow full of ripening pumpkins, harvested before a cold stretch that would have damaged them.

Eat The Weeds

"What is a weed? A plant whose virtues have not yet been discovered."

- Ralph Waldo Emerson

Dandelion salad greens, plantain leaf chips, roasted burdock root, blackberry leaf tea. We have so much more food on our little acreage than what grows in our tended garden beds. And the best part about it? I don't have to sow, water, or weed. I simply harvest the bountiful weeds at the peak of their season.

The first spring we lived on our little homestead, I discovered a small stinging nettle patch behind the chicken coop. It was sheltered from the sun until afternoon, and the earth stayed damp most of the day. I may have let out a squeal upon discovery. It's not that I hadn't seen nettle patches in ditches or along railroad tracks sprouting up in the weeks leading to this discovery; it was that I now had my own nettle patch on my own property that I was so excited about. I knew that there had been no harmful chemicals sprayed on our property where they were growing, so essentially I had organic nettles at my fingertips.

I watched the stalks grow daily, until they had reached about eighteen inches tall. Their distinct hairs along the stem and underside of the ovate, toothy leaves was an obvious reminder of the formic acid that causes a stinging rash on the skin when handled carelessly; a sting of which I was very aware from my childhood days. On the next sunny morning, after the dew had dried, I made my way over to the nettle patch with a wicker basket and my kitchen shears. Bare-handed, I gingerly snipped the tender, young leaf buds from each stalk, and let them fall directly into the basket.

It is often said that nettles will sting when you've gathered enough, and even though I am aware of this warning, I often find this to be the case. Typically as my basket fills, I tend to become less mindful of the task at hand, and I brush against the delicate, stinging hairs of the nettle in the process. It is as if the plant is giving a gentle reminder not to harvest more than is needed. As I gathered up my basket and shears and headed back to the house, I couldn't help but thank the Creator of all of these weeds. The earth is so full of nourishing food, if only we would take the time to look under our feet and appreciate it all.

The most common way I use my freshly harvested nettles is to infuse them in raw apple cider vinegar. It takes about six weeks of soaking for all of the nutrients to pass from the leaves to the vinegar, while shaking the jar every day or two, to keep rotating the leaves at the top of the jar so they stay submerged. The infusion makes the most unexpectedly delicious vinegar I have tasted—its

Stinging nettle, carefully harvested with scissors, so as not to get stung. Half of the leaves will be dried; the rest will be infused in apple cider vinegar.

flavor is much like strawberries and balsamic. I also use the nettle greens from the infusion to make a lovely pesto.

An exciting fact I also discovered that first year is that nettles produce a second, although somewhat smaller crop in the fall. I was able to return to my nettle patch to harvest enough leaves for another small batch of vinegar and leaves to dry for winter herbal tea blends. The humble nettle is chock-full of iron, and vitamins C and K. It acts as a diuretic, aids in balancing blood sugar, and can eliminate uric acid from joints, helping those with rheumatoid arthritis. I simply cannot get enough of this wonderfully nutritious weed. By spring, I have usually run out of the vinegar, pesto, and dried leaves I have put away and cannot wait to harvest yet another crop.

Dandelions are one of those tenacious weeds that has made its way virtually all over the globe. Thankfully so, as the flower, leaf and root are all packed with nutrition. When we lived in the suburbs, we were surrounded by perfectly manicured lawns that hardly showed a sunny yellow flower, even in their peak spring season. Spend any amount of time outside on a weekend in suburbia, and you quickly discover why—homeowners are mercilessly poisoning or digging up these little yellow lawn monsters. It is such a shame that in a few short decades, at least in North America, we have forgotten the value of weeds such as dandelion.

Dandelions, like nettle, produce both a spring and fall crop—spring for the tender leaves and flowers, and fall for the flowers and roots. I practically run outside doing cartwheels at that sign of the first dandelions emerging from the cold, hard earth. They are one of the first indications that spring is on its way for real—and one of the first spring foods for the bees. I nibble on a raw leaf or two as often as I am outside; its bitter bite on my tongue awakens my senses much the same way the way the crisp air of early spring does when it nips at my cheeks.

I find it an absolute delight to watch a graceful honey bee dancing among the vibrant yellow flowers. Daily, more and more of these humble perennials erupt into tawny blossoms, spreading across the carpet of emerald colored grass. I ask Mr. Green Thumb to hold off on mowing the lawn for a week or two until the bees and I have had our fill, and he nods in agreement. But it is torture to watch these enchanting posies wax and wane, while

The humble dandelion, considered more of a nuisance than of value, grows freely in our fields and lawn; its petals, leaves and roots are all useful around the homestead.

waiting until there are enough other pollen sources to sustain the bees before I begin my dandelion flower harvest.

On a clear, dry day after the dew has dried, I head outside with a bowl to collect the tiny petals of this ubiquitous weed. With all five fingers, I grasp the tips of the petals, and pluck—a slight pop can be heard as they release from the base of the flower. My youngest, CJ, is working beside me, determined to continue as long as I do. This goes on for some time, bent over at the waist plucking up cluster after cluster, until my back is aching and my fingers are stained saffron. Standing up to stretch and survey the damage, I can hardly tell I've taken any, and yet my bowl is already half full.

"Just a few more," I whisper to CJ and the cat who has wandered over to see what all the fuss is about and is now rubbing up against my legs; the bees are wobbling all around us, to and fro, weighed down by the granules of pollen stuffed in their pollen baskets.

There is no sense in drying the myriad petals I have collected— we all know what happens when the flowers dry up—POOF, an army of winged flower seeds are at the ready. I simply let them wilt for an hour or two, allowing the creepy crawlies time to evacuate. The petals then head straight for the mason jars I have lined up. One jar is destined to become dandelion petal infused olive oil for salves and lotions; several others will become rustic dandelion wine, and yet another will become dandelion flower infused vinegar. "Let me help, Mommy," CJ says as she reaches for the bowl, ready to stuff the jars full with her chubby, cherub hands.

And what is there to say about the modest plantain leaf, which I first discovered in the rhubarb patch? While the broadleaf plantain works wonders for any sort of external skin issues, the closely related lance leaf, or narrow leaf plantain is incredibly soothing for all sorts of internal digestive ailments. Simply chewing up one of these leaves daily for a week will leave the stomach settled and the large and small intestines working better than before. Lance leaf plantain has settled many an upset stomach at our house, and sooths any sort of bloating and pain from eating too much of a food one is sensitive to, or too many treats around the holidays. Best of all, plantain grows right in the lawn for over half of the

The rosehips, kissed with frost, are now ready to harvest. They will be boiled in water, and the infused liquid used in my elderberry syrup.

year. I never collect too much to dry, as it is readily available for so much of the year.

And then there are rosehips. The ruby red bulbous fruits of the rose bush are more than a splash of alluring red in the thick of winter; they are loaded with vitamin C and are one of the star ingredients in my elderberry syrup—a healthful syrup I make to help keep fall and winter colds and flus at bay. It also makes a subtly fragrant tea which can be drunk throughout the cold months. While the cultivated roses also produce edible rose hips, the wild roses produce the most nutrition. If you take the time to scrape the seeds out of each rose hip, the fruits will also make a thick, stewed jam equally full of all of the nutrition of a long steeped tea. While I have yet to take the time to do this, I may employ my growing sprouts to do this for me this coming winter, when we harvest this year's rosehips.

Blackberry is another wild weed that is rich in vitamins—and not just the fruit either; both the leaves and roots are also worthy of a place in the herbal cupboard. The leaves, much like raspberry leaves, are astringent and soothing for an upset digestive system. They also help to tone the stomach and other smooth internal muscles, including the uterus—which is why raspberry leaf tea is so often recommended for women to drink in their third trimester of pregnancy. Blackberry is equally as beneficial as raspberry leaves in this regard. The roots also have a place in my medicine cabinet. They are great when made into a tincture—plant material that is infused into alcohol or glycerin—to extract their medicinal properties. Around these parts, this tincture helps to ease the discomfort of an unhappy bowel during a bought of the flu.

I love many more wild and weedy medicinals, and I anticipate even more that I have yet to learn of, but these above are a collection of my favorites. Best of all, they all grow on and around our homestead.

Stinging Nettle Infused Apple Cider Vinegar

Ingredients:
1 quart (litre) raw (unpasteurized) apple cider vinegar
1 large bunch freshly harvested stinging nettle

Instructions:
• Using a kitchen or garden glove, firmly shake out the leaves to remove any insects or debris that might be on the leaves

• Allow the leaves to wilt for a couple of hours, or overnight, in an out of the way place, out of direct sunlight

• Using gloves or tongs, stuff the leaves into a quart sized jar and cover them with apple cider vinegar.

• Using a stainless steel utensil, swirl the leaves around ensuring that they are all covered and that there are no air bubbles trapped.

• Place a small square of parchment or wax paper between the jar top and the lid to prevent the acid in the vinegar from causing the lid and band to rust.

• Set the jar in an out of the way place, or in a cupboard, out of direct sunlight for 4-6 weeks.

• Strain the nettle leaves well, and store the vinegar in a sealed jar. Use in marinades and salad dressings, or mix 1 teaspoon into a glass of water, and drink daily.

Poultry Menagerie

"... the Truth the Turkey is ... a true original Native of America.... He is besides, though a little vain & silly, a Bird of Courage."

-Benjamin Franklin

Last spring, Mr. Green Thumb decided it would be fun to raise turkeys. "We'll have our own turkey for Thanksgiving dinner," he said. My ego got the better of me as I envisioned setting down a platter of our home-grown turkey as the centerpiece for our festive feast. Even though I knew I was adding another task to my to-do list, I agreed.

We picked up an extra-large cardboard box from a big box chain in town. Filled with wood shavings, and with a heat lamp dangling from a hockey stick and secured by zap-straps, we had ourselves a turkey poult nursery. The eight baby turkeys we had also picked up that day huddled together under the light and settled in for their first night in our garage. I was practically counting down the days until Thanksgiving.

Within that first week, one of the poults died unexpectedly. It was lying in the corner of the box, stiff and lifeless when the kids went out to feed them before school. They weren't as bothered by this death as they used to be, although we did still hold a little ceremony for the turkey poult when they got home that day. Otherwise, I didn't think much of its death—hatching and raising chicks had proved the same outcome; the weak birds often die early on in life, which we have come to accept, and expect. In this case though, it was a foreshadowing of things to come.

The poults ate and grew, and ate and grew some more, and before we knew it, they were flying up out of their nursery box in order to explore our garage. It was time to find them a larger home. We moved them into our portable chicken tractor, laid down a fresh layer of bedding, set up their food and water, and let them settle into their new surroundings. They really seemed to love the extra space to run and explore. One morning, shortly after the move, we found another one of them dead on the ground.

The turkey toms started to outpace the hens in size, and we were excited at the prospect of a large Thanksgiving bird. Within weeks of moving in, the poults were actually starting to outgrow the chicken tractor—at just several months they were already the size of our large hens.

As I fed the birds that morning, I decided that once the kids were home from school, we would each carry one or two poults up to the chicken coop. That would give the birds ample room to stretch their legs and roam around. However, when the kids and I walked down to the coop that afternoon, only five poults could be seen—one of our big boys had somehow drowned himself in the water bucket.

A hen scratches the dirt, looking for juicy bugs, in front of a blooming rosemary bush in one of my herb beds.

Shortly after our poults had been upgraded to the big coop, one of the hens disappeared with nothing more than a pile of feathers left in the neighbor's field adjacent to the coop. She had cleverly escaped, only to have her glimmer of freedom snatched away in the jaws of a hungry coyote or raccoon. With only two smaller hens and two toms left, we knew we needed to protect these four birds—the future of our Thanksgiving was at stake.

That evening, once we were finished supper, and the birds were all on their roosts for the night, Mr. Green Thumb and I crept into the coop. One by one we lunged toward the turkeys, taking several tries to catch each poult. In the process, we had riled up the coop into a frenzy. The hens flew up to the highest roosts, pacing back and forth, flapping their wings and squawking at us in an attempt to chase us away; but we were not deterred.

As Mr. Green Thumb held each turkey poult securely in his arms, I clipped the tips of the feathers off one wing and he released them back onto their roosts. We hoped this quick and painless process would keep the turkeys closer to home, and for a while, it seemed to work. They were no longer able to fly over the fence with their lopsided wings, and they appeared content to roam around the field the way the chickens did.

We noticed though, that after several weeks with the chickens, the turkey poults weren't putting on weight as fast as they had when they were on their own. Maybe the chickens were snacking on the turkey feed; maybe the turkey poults were filling up on chicken feed. Whatever the case, Thanksgiving was drawing closer, and we wanted our birds to bulk up a little more. And then, just when we thought we were in the home stretch, two more turkeys flew over the fence, disappearing with nothing but two piles of feathers. Apparently the predators weren't as picky as we were about the size of the birds.

The two remaining poults hadn't gotten as big as we were hoping, but the tom felt like he weighed around ten pounds—success in our books after all the struggles we had faced. The hen was smaller, still about the size of one of our larger hens, maybe around five pounds at best. But, we had finally reached the finish line, and they were ready to be processed.

We had another twenty cockerels and old laying hens to process as well, and we asked a fellow farmer if he could process our birds —he has an automated system for scalding and de-feathering the carcasses, something that can take hours to do by hand, but can be done in minutes with his equipment. Once we knew he had the time to help, we set a date and counted down the days. I was looking forward to adding these birds to the growing stockpile in our freezer of home-grown meat.

Unfortunately, that big beautiful tom never made it to butcher-day. Two days before we were scheduled to butcher them, I went out to feed the birds; I found his brown and cream colored feathers strewn around the grass on the other side of our fence—in the same field we had lost five other poults. I guess he craved freedom. Only one lonely turkey poult out of eight made it to that final day. She was not large enough for our Thanksgiving feast, but she fed our family well, and we were still thankful for the nourishment she provided.

Facebook farm pages are the worst. Farmers from all over the area try to sell beautiful farm creatures of every kind that other farmers, like me, are enticed into buying. It's because of a Facebook farm page that I almost ended up with a potbelly piglet this past Christmas! Emus, quails, every variety of chicken imaginable, ostriches, turkeys, ducks of all kinds, and peacocks—and that's just the birds. There are rabbits, goats, sheep, donkeys, cows, pigs, and horses for sale as well.

So when a local farmer who had become a friend, advertised one of her peacocks for sale, I couldn't resist. Our kids couldn't believe we were getting a peacock. "It's like we're a petting zoo," our seven year old said. "I can't wait to show my friends," replied our ten year old. They were so excited.

We all piled into the truck on a Saturday morning to pick him up. When we arrived, farm cats rubbed up against our legs, and we chatted about our animals and the hot summer we'd had. The kids ran to the swings, and began pumping their legs, trying to get higher in the air than their siblings. My friend's boys ran out into the yard at the sound of our sprouts' laughter and practically dragged my kiddos into their enclosed porch to see the newborn kittens.

The juvenile peacock was as beautiful as all get up. His sapphire blue neck gave way to a velvety brown and black saddle, and the tuft of delicate feathers on his head flashed a brilliant turquoise in the sunlight. She had him ready for transport in a large metal crate, and he darted back and forth, scattering his snack of sunflower seeds as he went. Mr. Green Thumb and I slowly lifted the crate into the back of his truck, trying not to aggravate him any further. We named him Mr. Peabody.

Once we were back on our farm we carried his crate into the chicken coop, just barely squeezing it through the doorway. My friend had suggested keeping him in the crate inside the coop for a week to help him adjust to his new home. The chickens rushed in after us, hoping for a midday snack. And when we fed Mr. Peabody some more sunflower seeds to try and help him settle, the hens clamored around the cage, stretching their necks in between the bars to get some.

I felt sorry for Mr. Peabody those first few weeks. His first week in the cage had him on constant fight or flight mode. Once we released him out with our other birds, he seemed happier but still incredibly skittish. We tried not to get too close so as not to unnerve him; he had enough trouble being bullied by the chickens. During feeding time, he hung around the periphery until most of the hens had eaten, waiting to take his turn. Although he was

Mr. Peabody, our handsome peacock, proudly perches on top of the gate to our orchard.

double their size, the hens would peck at him if he tried to eat at the same time as them.

These days, without a peahen in sight, Mr. Peabody seems to believe he is a chicken. The hens no longer intimidate him but include him in their feeding frenzy. Although his beautiful tail plumage has only started growing in—it doesn't fully grow in until peacocks are two years old—he has begun raising his stubby brown hind feathers and shaking his behind in his awkward teenage-style mating dance. The chickens ignore him, but that doesn't keep him from trying. His wings have fully grown in too, supporting his body, and they are able to lift him up to the roof of the coop quite gracefully. He spends much of his time up on top of the chicken coop relaxing and enjoying the view. If ever we lose sight of him, we can still hear his random, bellowing screeches and rest assured that he is nearby. When Mr. Peabody is not perched up high overlooking his territory, he can be found strutting around the homestead as eye-candy for any customers or friends that come by.

Since Mr. Green Thumb had insisted on turkeys, I insisted on ducks as well. What are a few more birds to add to the growing flock, I figured. We bought eight one-week-old Muscovy ducklings the same day we bought the eight turkey poults; for the first several weeks of their lives with us, they all lived together in the nursery box in our garage.

The morning after we bought them, my son and I went into the garage to feed the ducklings and change their water. We noticed that one little duckling had somehow gotten himself pinned behind the waterer in an upside down position away from the heat lamp, and had been like that for some time. His little body was already stiff, although his eyes were open and blinking. My son cradled him in his hands near to his chest as he carried him inside. I knew that the chance of this little one surviving was slim.

My son had a hard time leaving for church that morning. We set up a little hospital box in the laundry room, with shavings, water, food, and a heater. My son gently placed the duckling under the heater, and stroked his tiny body with one of his fingers. I saw a tear slip down his prepubescent cheek from where I stood at the doorway. "Time to go," I said, and he lifted the duckling to his chest once more to give him a hug goodbye.

The duckling was cold and stiff, sprawled out in the middle of the hospital box, when we got home. There was nothing more we could have done, and my sweet sprout, who wears his heart on

CJ cuddling with her hen Dotty.

his sleeve, ran up to his bedroom as large crocodile tears streamed down his face. When he came back downstairs about an hour later, he helped me bury the tiny duckling under the expansive branches of our ancient Magnolia tree—known to our family as the "chicken tree" because of all of the chickens we have laid to rest under its branches.

At about the same time the turkeys started to fly out of the nursery box and explore, the ducklings discovered how fun it was to completely upend their water dishes, creating a soupy mess of feces-filled shavings along the bottom of most of the box. It was time for them to move. The turkey poults had already been moved into the chicken tractor, and the ducklings were still too small to mingle with the chickens, so we set up a temporary pen for them using construction panels to fence them in, and one of our kids' old Fisher Price play houses for their shelter. We also gave them a large water trough so that they could splash to their hearts' content. And splash they did.

I was amazed those first few days to find that their trough was only half full, just hours after I had filled it to the very top. On cooler days, the ducklings would just dip their heads into the water while standing on the edge of the trough, and give a little shake before waddling away; on warmer days, they would dive right into the water, splashing and playing together. They bobbed under the surface, and then popped up again, flapping their wings and sending water droplets every which way.

The survival rate of the ducks wasn't much better than that of the turkeys. Only four of the eight made it to teenage-hood, and only one of those was female. I had my hopes set on a couple of breeding pairs, but one would have to do. And since the other two males would have been lonely and might have tried to mate with and injure our hens, they ended up in the freezer as well.

As the Muscovy pair continued to mature, the male developed the red, knobby, cartilage-like protrusions all over his head, called caruncles, which are characteristic of these ducks. He is quite a sight to behold. His slick, black feathers shine opalescent greens and purples in the sunlight, which contrast with his brilliantly red head. He has a majestic air about him as he waddles about the property.

His companion disappeared without a trace in late winter. At first we assumed she was sitting on a clutch of eggs somewhere, and we just hadn't found her yet. But as the days turned to weeks,

One of our muscovy hens with her brood of ducklings. Muscovy hens make excellent mothers, sitting on their fertilized eggs and tending to their ducklings.

and one month turned into two, we knew that she wouldn't be coming back. In the months that he'd been without a partner, our male Muscovy was subdued. Instead of strutting around the homestead, he sat mostly in one place, either in one of our garden beds, or on the edge of a planter pot we have up against the coop. He was lonely and needed a new friend.

We brought home a trio of young female Muscovys a few months later. Our drake immediately took to them, and the foursome could be found waddling around the property together, our drake honking and calling to his new friends whenever they stepped more than a few feet away. We named him Quack the Duck, because he didn't stop calling out to his female companions for weeks after they arrived. Fortunately, they are as equally enamored with him, and the three of them happily follow Quack wherever he goes.

Sauerkraut

"There is an underlying unity to all vegetable fermentation: By keeping vegetables submerged under liquid, you create a selective environment where molds and other oxygen-dependent organisms cannot grow, thereby encouraging acidifying bacteria. Beyond this simple technique...vegetable fermentation can be quite varied and quirky."

-Sandor Ellix Katz

It took years for me to work up the courage to ferment food. And after I had become brave enough to make it, I waited longer still to trust that it was edible. Even after I had eaten store bought varieties, and pored over books and websites explaining the step by step process, inherent safety, incredible health benefits, and long history that fermentation has had, I couldn't wrap my head around the idea of consuming food that had been left on the counter for, essentially, controlled decomposition. I made several batches of sauerkraut only to throw them into the compost heap without tasting them.

There was no pivotal turning point that changed my mind; only that enough time had passed, and the idea of fermentation had fermenteded in my mind long enough, that I felt like it was time—time to taste my homegrown, home-fermented beast and face my fate. I did so with a very miniscule, limp strip of cabbage. If I'm not sick by tomorrow, I thought, I'll eat a little bit more. And so I did. Every day I ate a little more sauerkraut, until I had worked up to eating an entire fork-full from the jar. It was fabulous; and it was liberating. This thing that had caused so much trepidation had been conquered. My journey with fermentation had begun.

And so I began experimenting with different types of vegetables, fruits, and flavors in my ferments. I pickled cucumbers with chilis; grape tomatoes with garlic; apples with cinnamon sticks; blueberries with honey; green beans with dill; carrots with ginger; beets with corriander. I couldn't get enough. Jars at various stages of fermentation lined my countertops for weeks at a time. I branched out to kefir and kombucha and was continuously trying to feed my family all of the concoctions I had made. They were less enthusiastic about it than I was, although I do believe I have created a bit of an addiction to kombucha in my youngest, CJ. We enjoy sipping a little cupful together most mornings.

Last year, after watching the cabbage grow in the garden all summer, I harvested it and turned all five bulbous heads into the largest batch of sauerkraut I had made to date. An abundance of quart and half gallon mason jars were stuffed with exceedingly fresh, julienned cabbage that had been in my garden only hours earlier. After I topped each jar with a simple brine of water and salt, I pressed a water glass into the top of each opening to keep

Sauerkraut

A Cabbage grown in our garden, destined for a batch of sauerkraut.

the brine above the cabbage and tied a dish cloth around the rim of each jar. Now they would sit, undisturbed and out of direct sunlight, for at least a month, allowing all of the enzymes and bacteria naturally found on all fruits and vegetables to get to know each other and create their magic.

Day by day, little changes appeared in each jar. Twenty-four hours after the jars had been filled with cabbage and brine, little bubbles could already be seen rising to the surface of the liquid;

at three days, the bubbles were larger, and more active, and the bright green of the cabbage had become more subdued. Daily, I peeked under the towels to check on my living food; daily, I felt amazed at the alchemy taking place before my eyes. At around a week, I sampled the sauerkraut—the salty and tangy flavors mingled on my tongue, while the crisp crunch of the cabbage created a nice mouthfeel when I bit into it. Tasting my homegrown, homemade, healthful concoction was immensely satisfying. At one month, pleased with the sourness of the 'kraut and knowing that the flourishing probiotics were at their peak, I tightly capped each jar and placed them into the back of the fridge to bring the fermentation process nearly to a halt. Each jar would keep well past six months in cold storage if we didn't eat their contents first.

Not long after my journey with sauerkraut began, I visited some family friends and brought them a jar of my humble 'kraut as a gift. Having German heritage, I knew their taste buds would offer a fair critique of the flavors that had developed in my kitchen. To my delight, they raved about the authentic flavor and crisp texture of the cabbage, and I happily jotted down for them the basic recipe I had used. As I was gathering my things to leave, my friend told me that she had an old crock sitting in her basement gathering dust. She asked me if I wanted it. It had belonged to her parents, and she remembered when she was a child them making sauerkraut in it, although she hadn't carried on the tradition herself. I couldn't believe my good fortune as she handed me the two gallon, ceramic crock. I felt honored to receive this historic treasure. Whether full or empty, this crock has a prominent place near our kitchen table and is a continual reminder of good food, good friends, and the joy that comes from combining them both.

A jar of newly made sauerkraut rests on a plate. As the cabbage begins to ferment, the liquid bubbles up and spills out of the jar.

Basic (Unquirky) Sauerkraut

Ingredients:

1 quart wide-mouth mason (or other canning) jar

1 small water glass that fits inside the mouth of the jar

1 small (one pound) green or red cabbage

3-3 ½ cups filtered or well water

3-3 ½ teaspoons non-iodized salt or sea salt

1 teaspoon caraway seeds, or 2 teaspoons dried juniper berries (optional)

Instructions:

• Clean the quart jar in the dishwasher or with very hot, soapy water to sanitize. Find a glass that nests inside the opening of the jar.

• Very thinly slice the cabbage with either a mandolin or chef's knife; using a food processor will make the pieces so small that the final result can be quite mushy.

• Mix one cup warm water with the salt; then add the remaining two or two and a half cups cool water to bring the brine to about room temperature.

• Layer the cabbage and the brine, and sprinkle a pinch of caraway or juniper, about a quarter of the jar at a time, pressing down each layer of cabbage with a stainless steel spoon or other stainless steel utensil.

• Once the cabbage comes up to the shoulder of the jar, pour the brine up to the same height, and then nestle the glass into the cabbage and brine; the brine should rise above the cabbage, while the glass holds the cabbage below the brine.

• Place a plate under the jar, if desired, as the brine will often leak out of the jar.

• Tie or place a clean dish towel over the jar and place it out of direct sunlight.

Sauerkraut

- Taste the cabbage at one week, and again every week or so, until the flavor tastes just right. Keep in mind that the probiotics are at their peak development at around one month to six weeks.

- Once fermentation is finished, remove the glass, screw on the lid, and place in the fridge to enjoy for up to six months.

An Unlikely Farm Dog

"Until one has loved an animal, a part of one's soul remains unawakened."

–Anatole France

The day started out like any other—outdoor chores with our tiny, yappy Yorkie named Scarlett by my side. As I moved from the chickens to the pigs to the goats, she faithfully followed me. Two years into our routine, it had practically become a synchronized dance. She bounced around my feet as I walked up to the chicken run; she stood guard while I opened the gate, ensuring no hens would try to dash out while I fed them. From there we moved to the outside of the run, gently herding any escapees back inside for their morning meal.

Next, we walked down to the pigs and goats, and she dashed ahead of me, cheerfully barking at the animals to let them know that breakfast was on its way. The goats and pigs snorted and grunted at her in mild annoyance—maybe because she was too chipper, or maybe because they were hungry. Either way, once the food arrived, the goats got to munching and the pigs got to gorging, and they paid no attention to their pint-sized pal. I don't know if it was a jogger or a car, or even another dog that caught her attention, but it was then that she took off and went straight up the driveway to the road to let any would-be-trespassers know who ruled our roost.

And just like that, she was gone.

We had welcomed our sweet little pup to the family one beautiful, sunny October day, about a year before we moved to our homestead. Our oldest, who was eight at the time, chose her over her litter mates because she had a prominent tuft of orange hair on her nose, which he thought was simply adorable. He also insisted that we name her Scarlett. Since the other kids had no objections, that is what we named her.

She captured our hearts right from the beginning. She loved to be cuddled and held, and followed the kids and me around wherever we went. She even tolerated being picked up by our youngest, who was a toddling one-year-old at the time. Scarlett had her quirks, like barking uncontrollably at windshield wipers and flapping flags, but her loveable personality more than made up for it. Aside from the rainy days to avoid her inevitable windshield wiper serenade, she came with me almost everywhere I went.

When we moved to our homestead, we weren't sure how she would adjust. Outside, she had only ever been in an enclosed backyard or on a leash. Now, she would have room to run free. At first we kept her on a leash and then only let her outside when one of us could supervise her. Eventually though, we felt comfortable enough letting her out for short periods of time completely on her own. Right from the beginning, she acted like she owned the

Our sweet farm pup, Scarlett, just days before she disappeared.

place. Her tiny six pound frame dashed up the driveway and back down to the house whenever a car drove past. She trotted along the perimeter fence and often verbally defended her home turf when she came face to face with the neighbor's cows along the fence line.

When we brought home chickens several months after moving, she was incredibly curious and liked to circle their chicken tractor, barking at them to get a reaction. They flapped around inside the safety of their enclosure and even hopped up onto the roosting post where they were out of Scarlett's sight. Her curiosity didn't cease when we built a larger chicken coop and run and added another thirty-five birds. She would run up and down along the fence, stopping to bark at any chicken that dared come close to her. But it was always in fun. She was surprisingly gentle with the birds that jumped over the fence. Although she chased them, she never tried to catch them or touch them.

Eventually, she actually learned to herd the chickens with me when they escaped. After I went inside the chicken run to feed them, I would open the gate and walk along the outside of the fence, encouraging any birds that had hopped over to walk back in the direction of the entrance. Scarlett would bounce alongside them, directing them to the opening, and barking at them as necessary to get them moving. Although many of our heritage birds were larger than she was, they knew she meant business and didn't hesitate to follow her lead.

She accompanied me on my other farm chores as well. She loved to rush ahead toward the pigs as I carried the heavy buckets of feed behind her. They were the endless recipients of her playful barking, and they often stuck their snouts right through the fence and snorted back at her. I'm sure they had many a conversation that I wasn't privy to. Once, Scarlett discovered a large enough gap under the fence to squeeze through and found herself face to snout with two very curious pigs. All at once she seemed a lot smaller, and her bark took on a more panicked tone. On their side of the fence she was the foreigner, no longer in control, and she knew it. As the rest of the pigs lumbered toward her, she dashed for the opening and wriggled her way back to safety just as fast as she could.

As I was feeding the chickens, almost a year after Scarlett disappeared, a ball of black fluff caught my eye scratching in one of my herb gardens. My heart skipped a beat. Could it be Scarlett? Had she found her way home? I found myself holding my breath as I slowly moved forward to examine the scene. When I realized it was just one of our black Australorp hens searching for grubs, I felt as if I had had the wind knocked out of me. Until that moment, I wasn't aware that I had been subconsciously hoping she would come back to us one day. It was the fact that we never knew what became of her that made it so hard to accept. For almost a year, the sprouts and I coped with the loss by making up stories of Scarlett being found by a happy, loving family who treated her like a princess. I could not let myself imagine a worse fate.

One year had passed since Scarlett had disappeared, and we decided it was time for a new farm pup. Our youngest had just turned five and diapers were already a distant memory. I felt that our family was ready to love another dog, and I was ready to take on the challenge of housetraining a puppy. We surprised the sprouts with a little ball of black fur we named Roxie.

Roxie did not, however, have the gentleness that her predecessor Scarlett had had; at only six months old, and no larger than the chickens we kept, she escaped from the house without her leash on, and proceeded to break the neck of one of our prized blue-egg laying hens. Twice more that week, she got out of the house, and went straight for a chicken's jugular each time, shaking the chicken in her mouth back and forth until it was limp and lifeless. We knew that her instincts were too strong to stay on as a farm dog; she would need to be rehomed. Although she was loveable, even our sprouts knew that for the safety of our hens, she needed to go.

My dream of having a well mannered, family dog who knew her way around farm animals was all but gone. It had been a year since Roxie had moved on to live with my in-laws. The sprouts loved visiting her and cuddling with her when she let them, but they were starting to get doggy-fever again; they started by asking if we would just consider getting a dog, even if it wasn't right away. Then, they suggested we could plan to get one for Christmas, which would give us lots of time to prepare. One day, our second oldest sprout looked up dogs for sale on Craigslist, and the sprouts all went gaga over a middle-aged Chihuahua with a complicated past.

Not long after, our oldest sprout prepared a presentation on his school tablet, including all of the reasons that kids should have a dog. His research included appropriate dog breeds; physical, mental and emotional benefits based on scientific studies; the responsibilities that kids develop from owning and caring for a dog; and photos of the some of the cutest dogs I have ever seen. He had been busy typing away one rainy Saturday, and I assumed he was working on homework. When he called a family meeting in the living room later that afternoon, I learned otherwise. He had been working on a slick Powerpoint presentation to try to persuade us to get yet another dog. It worked.

Behind closed doors, Mr. Green Thumb and I began to discuss the possibility of trying once more to find a farm pup. When some friends of ours let us know they had one last puppy, named Daisy, available out of their most recent litter of miniature Australian Shepherds, we decided to meet her. From the moment we walked in the door, she acted as if she belonged to us. She rolled onto her back so that our sprouts could take turns giving her belly rubs, and she hardly left my side all evening; it certainly seemed as though she had chosen us. Although we hadn't planned for it, much to the excitement of myself and the sprouts, Daisy came home with us that night.

Our lively farm pup, Daisy. Her endless energy has been challenging to keep up with, but she is settling in nicely.

Daisy has contentedly fallen into her role as farm dog. She loves running around the homestead, barking at songbirds, wild rabbits, our animals, and the kiddos as she passes. Her nose is instinctually curious, and she cannot help but sniff out every smell she encounters, whether it is flower, tree, fence post, shoe, or compost bin. She is still young, but her herding instincts are strong, and she loves to boss around the hens by chasing any that have hopped the fence. She and I equally love each other's companionship, and she is just as happy to curl up against me as I write, as she is bounding through the field while we complete the chores. The sprouts adore her too, and I am only slightly envious of them taking turns sleeping with her in their rooms for the night.

Fresh Baked Bread

"Where there is no bread, there is no life."

- Old German Saying

Something was almost always cooking when we arrived at my grandparents' house. On the days my mom took my brother and me over to visit after school, the aroma of fresh baked, whole wheat bread often greeted us. My mom's mom had mastered the art of bread baking. Loaf after loaf lined the countertop, cooling, while still more were baking in the oven. If there wasn't a jar of jam already open in the fridge, she would send one of us downstairs to her pantry, lined amply with jars of all sorts of fruits, pickles, jellies, and preserves. Blackberry jelly was my favorite, but that was reserved for holiday gatherings; the usuals were strawberry rhubarb, or apricot.

Waiting for the loaves to cool was torture, if they weren't already cooled enough by the time we arrived. I could not wait to sink my teeth into a warm, fluffy slice of bread, slathered with salty butter and a sweet dollop of jam. Both my brother and I ate slice after slice until our bellies were filled to the brim, and even then, it was tempting to indulge in another. Sometimes, we would still be visiting when my grandpa arrived home from work, and he would plunk down his metal lunch pail, and sit down at the table, while my grandma would serve him up a plate with several slices of bread. My brother and I would snicker to ourselves as my grandpa cut wedge after wedge of butter and set them atop his slice, and then bit into the bread and butter, leaving teeth marks in the butter.

On Sundays when we were growing up, my family, often along with my aunt, uncle and cousins, visited my dad's parents' house to share an early supper together, called faspa. The table would be spread with a variety of cheeses and meats, homemade jams and pickles, and of course homemade buns. My dad's mom had mastered homemade buns the way that my mom's mom had mastered bread; they were scrumptious. I remember enjoying many butter and pickle, or jam and cheese sandwiches on her freshly baked buns, much to the curious horror of the adults sitting around me.

For dessert, my grandmother would often serve dessert buns called kringel. Almost like a sweet pretzel, the dough was twisted together, then baked. Once the buns were out of the oven, she would drizzle the warm buns with an icing sugar glaze. And, as if we hadn't already consumed enough carbs in one sitting, these kringel buns were the perfect finish to the humble faspa meal. We kids were thrilled if our grandmother would offer to send some buns or kringel home with us for school lunches—my mom would normally not dream of sending us to school with unhealthy, white buns.

Fresh Baked Bread

A loaf of twenty-four hour bread, cooling. We can't wait to slice into it.

Once I had my own little family, bread and bun making were on the top of my list of skills to learn. I wanted my own sprouts to have fond memories of eating a variety of baked goods fresh from the oven, as I had done as a child. Bread baking came easily. Flour, water, salt, yeast, and a pinch of sugar to kickstart the yeast. I quickly traded in my electric dough hook to complete the task by hand. I found folding and working the dough to be cathartic. I was proud of the golden brown loaves I pulled out of the oven, made with my own two hands.

My first attempt at buns was not as successful. I had used a recipe I had received from my auntie, and they had turned out like golf balls. I couldn't figure out what had gone wrong, as I had followed the recipe perfectly. As I was describing my method to my auntie the next time I saw her, she pointed out that I should not have punched down the dough after letting it rise—something I assumed was the same as with bread baking. The next time I made them, they came out perfectly. I knew that my grandmother would have been incredibly proud of me for providing for my family in this way.

I am not sure what motivated him, but shortly after I started baking, Mr. Green Thumb wanted in on the action. He had recently come across an article describing an easy method of bread baking, called twenty-four hour bread. When I read the recipe and instructions for the first time, I laughed aloud. The dough was not meant to come together into a round ball, but instead, it was to look like "a shaggy mess" in the bowl. Sounds about right for us, I joked. He made the bread, and it was divine.

With a newly discovered appreciation for baking, Mr. Green Thumb became our chief baker. He experimented with different ratios of rye, spelt, white and whole wheat flours in the twenty-four hour bread, as well as with recipes for sandwich and artisan loaves as well. His love of bread baking must have reached the North Pole because he received a bread-making machine for Christmas that December. And, it was a good thing too, because after months of health complaints and doctor's appointments, I was diagnosed as gluten-intolerant just two months later and could not have brought myself to bake any bread that I could no longer eat.

At first, I resented Mr. Green Thumb for continuing to bake bread when I couldn't eat any, even though he was providing for our family in this way. My journey of grieving over something I could no longer have lasted a long time before I worked up the courage to attempt baking my own gluten-free loaves next to his wheat ones. Now, when he pulls out the glass bowls to bake, he takes one out for me as well. He and our girlies mix the flour, water, yeast, sugar, and salt into a shaggy dough, while I mix my gluten-free sourdough starter into my pre-measured gluten-free flours. We then set the bowls aside, giving both doughs time to proof overnight.

Nestled into the covered dutch ovens, both the gluten-free and wheat loaves slide into a four hundred degree oven—Mr. Green Thumb's for about an hour and fifteen minutes, mine for nearly two hours. We remove the lids of the dutch ovens for about the last twenty minutes of baking time to ensure a crisp, golden crust atop our artisan style loaves. While the wheat loaf has risen substantially during baking, the gluten-free loaf stays flatter, but the results are not disappointing. There are little air bubbles throughout the finished loaf, and the bread has a satisfyingly sour tang, along with a good, spongy texture.

With a flavor and mouthfeel this pleasant in a loaf of gluten-free bread, I don't feel like I am missing out these days. We pop the next batch of wheat loaves in the dutch ovens and back into the oven to bake. The sprouts are hovering around the first loaf, dancing around the kitchen full of life and excitement, willing the bread to cool. Finally, one of us slices into it, and they each smear their slice with butter, and maybe a scoopful of blackberry or grape jelly as well. In our house, gathering our family around a loaf of freshly baked bread feels like a special enough occasion to indulge in the jelly.

Grow. Cook. Eat. Share.

Twenty-four Hour Bread
(Adapted from Jim Lahey's No Knead Bread recipe)

Ingredients:

3 cups flour (white; half white/ half whole wheat; spelt— we have used all of these options and they all work well)

1 1/2 cups warm water

1 1/2 teaspoons salt

1/2 teaspoon instant yeast

Instructions:

- Measure out flour; add salt and yeast. Stir to combine.

- Add the water, and stir until it comes together but is still sticky.

- Cover the bowl; let it sit for up to 24 hours in a warm area.

- Dump the dough onto parchment paper, using a spatula to scrape it out if it sticks to the sides; fold up four corners of the dough on top of itself. Let it sit for an hour.

- Pre-heat the oven to 400F. Preheat a large oven proof stock pot or roasting pan with a lid for at least 15 minutes.

- Lift up the dough by the parchment paper and place both parchment and dough into the preheated pot; bake with the lid on for 30 minutes.

- Take the lid off and bake for another 15 plus minutes. The best way to be sure that the bread is cooked through is to use a meat thermometer to check that the internal temperature of the bread is 190F, or 185F for high altitudes.
- Allow to cool for about one hour before slicing into the bread.

Bucks will be Bucks

"It's good for a person to be terrorized by a goat. Hard to get high and mighty when there's something chasing you for vegetables."

- Natasha Pulley

Our first buck, Rowan, came to the farm not long after we had purchased our first doe, Clover. He had a beautiful, buckskin colored coat—a caramel brown body with black hair around his neck and down his nose, and black hooves. From day one, he was friendly, and as calm as could be. He loved to be cuddled and ate grain straight from my and the sprouts' hands. At three months old, he was the perfect gentleman.

Rowan came to our farm along with his castrated twin brother, Gingko, to keep him company. Gingko was much more timid—he ran away from us whenever we entered their paddock. We often sat down in the grass, trying to keep as still as possible in order to coax him to come over and say "Hi." We would also bring along a bit of barley and gently shake it in the bucket to entice him. Occasionally his curiosity got the better of him, and he would come up to us and give our arms or hands a nudge before burying his nose into his treats.

Rowan hit puberty at around six months, and his gentle mannerisms gave way to loud, demanding teenaged antics—the actions that bucks are so often described to have. As cute as he was, and as much as I loved him, he had begun to turn into a typical hairy, smelly, noisy buck. Charged with testosterone, he was raring to do the job we had bought him for. Regularly, he would mount Gingko, who quickly tried to escape. We knew that it was time for Rowan and Clover to meet.

We had been building a permanent goat shelter that fall, and Clover and Acorn had already been living in it for a couple of weeks. We moved Rowan and Gingko in on the day that their shelter was complete—the day the first snow fell that season. Our little herd was together for the first time, and we were thrilled at the prospect of little kids running around come spring. Clover and Acorn were less enthusiastic to welcome Rowan and Gingko into their home. It didn't help matters that Rowan now tried to mount all three of his herd mates.

Clover was as unimpressed with Rowan's advances now as she had been when he was on the other side of the fence. She raced to the other side of the pasture whenever he got near her. He only got noisier and smellier. As bucks often do when trying to attract a doe, Rowan regularly peed all over his front legs, hoping his fresh, musty scent would entice his new friend. He also rubbed his head on the fence posts, releasing a particularly strong goaty odor through a gland at the top of his head.

Although it was most certainly not love at first sight, Clover was eventually won over by his charming ways, and soon enough, goat kids were on their way. In late winter we bought two more does, Poppy and Violet, bringing our total to six. They weren't

Our buck, Hawthorn, asserting his dominance by bravely standing on the hill that usually only our buck Rowan stands on.

nearly as turned off by Rowan's tactics and were happy to get to know him. We were hopeful that all three does would give birth the following spring.

As I've mentioned, bucks can be dirty, stinky boys, especially while they are in rut. I was reminded of this recently on a coffee date with a friend that I hadn't seen in a while. I had rushed out to feed the animals before I left, including our smelly boys. While I had the good sense to change my top before I left, I figured a quick wipe-down of my jeans would be good enough. Off I went to the chic coffee shop where we had planned to meet.

As I was driving, I caught whiff of a strong goaty scent wafting up at me. I didn't think much of it until my friend and I sat down with our coffees. She had a peculiar look on her face. It only took me a second to register what she was thinking before asking her if I smelled like a goat. She nodded, with a smirk on her face. Thankfully we have been friends long enough that she didn't think much of it either.

Clover was first to kid, giving us a pair of strapping young bucklings. One was the spitting image of his dad, and the other looked almost identical to Clover. Next was Poppy, producing darling twin doelings. The larger of the two was black with a splatter of white moon spots across her left side. The smaller one had a black cape down her neck, a white body, and a light brown hind end. Finally, Violet bore us a sweet doeling and buckling. The doeling looked exactly like her—light brown with a black stripe down her spine. Her little buckling was a mirror image of his dad, Rowan.

Out of the six kids born that spring, we decided to keep two doelings—the black one with white moon spots from Poppy, whom we named Lilac; and the light brown one from Violet, whom we named Lavender. We were thrilled that our herd had grown by an additional two does. But considering that Rowan had fathered our two youngest goats, we knew that we needed to find them their own buck if they were going to be bred the following spring.

We searched around for a buckling, but there weren't any available locally. We were too late in the season to find a boy for our girls. I had already asked to be put on a waitlist for the following spring, when a lovely buck with strong genetic lines came up for sale. A friend of Mr. Green Thumb's had tipped him off, and we immediately investigated. He was a handsome fellow—black with large white moon spots covering his entire body. He had a great frame and was a nice small size. We snapped him up at once.

The weekend after we bought him, we went to pick him up. The farmer had him in a small paddock with another buck. Mr. Green Thumb and I wheeled our large pet carrier on a dolly between us as we walked down the path toward him. The farmer jogged ahead and stopped at the barn to get a scoop of treats to help coax him into the carrier. But as the three of us entered his paddock, he turned on his heels and bolted. He was not in the mood to meet us or be captured. This was going to be interesting.

All in all, it required an hour of our time, dirt on the knees of our jeans, a belly full of goat treats, and our four sprouts bordering on anarchy in the truck. We had finally managed to round up our newest addition, Hawthorn the buck. He seemed much less elusive once in the crate, and contentedly nestled himself into the straw bedding to chew his cud. The whole ride home, the sprouts kept peering through the back of the pickup window to check on him; not once did he get up from his position. We hoped he would be as happy at his new home as he seemed to be in the back of the truck.

Hawthorn's quarters were with Rowan, Gingko and Acorn, on the other side of the fence from the ladies. Rowan didn't know what to think when Hawthorn emerged from the crate. According to Rowan, we had just introduced him to his arch rival. There was plenty of head-butting and aggression those first few weeks to sort out the new order of dominance.

As playful as it all was, Hawthorn was the outsider and he knew it. When I brought them their hay, Rowan would noisily munch away, flanked by Acorn on one side and Gingko on the other. One of them would inevitably butt into Hawthorn if he got close enough to take a mouthful. Often I saw poor Hawthorn foraging for leaves and twigs while the others got their fill. Only when they had had enough would Hawthorn timidly approach the feeder and help himself to the leftovers.

One morning I was running late for an appointment, and with efficiency in mind, I decided to feed the goats right before I left. Dressed in my city clothes, I proceeded to fill the females' feeder with hay. I then carried the males' portion over to their paddock and filled their feeder as well.

I had done a pretty good clean-up, brushing away all of the loose pieces of hay that had stuck to my sweater, or so I thought. It wasn't until I was midway through my appointment as I scratched the back of my head that I realized I had a very noticeable piece of hay clinging to my ponytail. I sheepishly pulled it out and tossed it in the trash can next to me. The two of us had a good laugh over that.

One day, shortly after we had brought Hawthorn home, I heard the most awful yelping noise coming from the bucks' enclosure. I ran out of the house and over to their paddock to check on them, thinking that one of our bucks must have gotten himself stuck or injured. It sounded as if one of them thought he was dying. I hoped against hope that Rowan hadn't done something to hurt Hawthorn.

All four boys looked at me when I arrived at the gate. They were happily chewing on the leaves of a sapling that had sprouted up in their paddock. There was no sign of distress anywhere. Even Hawthorn didn't appear to have sustained an injury. I turned to walk away, shaking my head, trying to figure out what had caused such a ruckus, when I heard the horrible holler again. If my sprouts hadn't all been at school, I would have mistaken it for the yelling of a preschooler being mercilessly antagonized by her older brother.

As I whipped around to look at the goats, who were staring back at me innocently enough, it dawned on me that this might in fact be Hawthorn, since I hadn't heard it before he arrived. And not only Hawthorn, but Hawthorn's normal voice. This screeching yell, alternating with guttural groans might in fact be the voice he was born with. And it also might have been part of the reason we got him for such a steal.

I wasn't convinced though, until it happened a third time, while I was standing there observing my new friend. It was him. And he wasn't in distress, being taunted by his peers, or caught in a fence. Hawthorn had simply become brave enough to let us hear his awful sounding voice.

Goats are naturally quite curious, and like toddlers, they tend to explore their surroundings by putting things in their mouths. On one such occasion, I was giving a friend a tour of our animals. As we stood next to the male goats, giving Rowan a good scratch on the head, Acorn proceeded to pull her beautifully ornate scarf through the wire fence, and gnaw on the tassels. I would have been mortified if she hadn't graciously laughed it off.

We have had to be more intentional about breeding this year, since we will be breeding two of Rowan's daughters as well. Gone are the days that the goats can all live happily under one roof for the winter—and we guestimate the due dates of our does. This year, Rowan will have dates with each of our three original does, but still be staying with our bucks in their shelter throughout the winter months. Once Rowan's daughters are old enough to breed, Hawthorn will be going on dates with each of them.

I've also become much more proficient at understanding the subtle and not so subtle cues that does give when trying to flirt with the bucks. Recently, Poppy was acting very different than usual. She had backed herself up against the fence between her and the bucks, her tail twitching ferociously. Rowan and Hawthorn were beside themselves, snorting, and whining in agony. With a quick check, I knew Poppy was in heat, and we needed to get her and Rowan into their own private quarters.

Once together, Poppy changed her strategy and ran from Rowan whenever he got close. But he wasn't the bull-headed animal I expected him to be in that moment. He was tender in the

Our buck Rowan, looking up from his lunch, wondering why I am bothering him.

way he pursued her. He would come up beside her and make a deep whining noise while gently pawing at her side with his front hoof. She would skittishly run ahead a few feet, and he would follow, repeating his actions. This went on for almost an hour before Poppy finally conceded. If things went as planned, she will give birth in approximately five months. Hopefully Clover and Violet won't be too long behind her.

I can't wait until our doelings are old enough for Hawthorn to go on dates with them. In the meantime, Hawthorn has also gone into rut, showing off for his pals by peeing on his front legs, rubbing his head on tree trunks and fence posts to release his musty scent, and curling his upper lip to show off his sexy front teeth. I'm sure the doelings will be impressed when the time finally comes for them to meet him.

SHARE.

Winter

Winter Solace

"What good is the warmth of summer,
without the cold of winter to give it sweetness."
—John Steinbeck

Without a doubt, Old Man Winter has been here. A sheath of ice has enrobed both evergreen and deciduous branches alike, and they twinkle in the early morning light. A thin layer of snow lies frozen over the vegetable garden, flower beds, and herbs, insulating the dormant roots from the harsh winds and freezing temperatures. The vibrant red rosehips, holly, and hawthorn berries are a stark contrast to the muted browns and grays of the season.

A hint of wood smoke hangs in the air as I go about my outdoor chores. Once again, the fog has rolled in off the Fraser River and it has settled into the valley, enveloping the fields of blueberry bushes and the farms that form a patchwork on the valley floor below us. We sit above the fog today, and I can see the faint glow of street lights flicker on and off on the adjacent mountain as the sun prepares to climb above its peak.

The air and the trees are motionless as I cautiously pick my steps down the uneven slope to the goats—quite the opposite from the howling winds and freezing rain we experienced just a week ago. Cedars, willows, maples, and firs all gave way under the intolerable burden of ice caused by last week's stormy weather. The tree line along the edge of our property is littered with their fallen branches.

This recent ice storm also left the area we live in without electricity for two days and two nights, as toppling branches pulled down any power lines in proximity. This past summer I had begun collecting water in miscellaneous bottles and jugs with this scenario and previous winter power outages still relatively fresh in my mind. Much to Mr. Green Thumb's dislike, these bottles began taking over the stairwell leading down to our basement—first along one side, and then the other.

Our oldest had recently moved downstairs and relayed his annoyance over having to walk past these bottles every day, and over the embarrassment that would ensue if his friends came over and saw them. They actually caused a bit of a kerfuffle the day that the delivery company brought our son's new bed downstairs to his room. I hadn't thought to move the water beforehand, and as the delivery person shifted his weight onto the first step, the headboard knocked over a bottle, creating a domino effect, which sent at least half a dozen containers careening down the stairs.

No one was frustrated or embarrassed though, the day the power went out and our well water stopped flowing. And I'll admit that for a moment I was relieved that the power actually had gone out, making all of my water saving efforts worth my time and energy, and worth the eye rolling that I had endured. One by one,

The snow beginning to fall, just after we finished building the goat barn.

we opened the gallon, half gallon and quart sized bottles and containers. Some went to flushing toilets, others for washing hands, and still more to fill our gravity-fed water purifier. I felt like we had been transported back in time—carrying water to cook and clean with; lighting a fire (on the gas stove) to boil water and cook our meals; and playing games and reading together each evening around the glow of the candlelight, albeit the battery powered type.

Those two days were actually a little magical. The kids were home from school, and Mr. Green Thumb was home from work. We spent our days lounging in pajamas, robed in blankets, with mugs of steaming cowboy coffee or hot chocolate at our sides. The kids set aside their typical bickering and pleasantly settled into a subdued state of relaxation. We took turns playing card games and board games while gusts of wind rattled the window panes, and the occasional cracking, falling branch interrupted the calm.

The animals didn't fare quite as well and kept huddled together in their shelters. We filled the goats' feeders with extra hay and barley, and gave the chickens extra feed to keep their energy and body temperatures up. As the de-icers had stopped working once the power went out, we also needed to repeatedly break up the ice in their water troughs. Unfortunately, one of our hens fell into the water trough after the evening chores, and I found her carcass the next morning, frozen into an inch-thick layer of ice on the surface of the water.

We lost a great number of animals to coyotes this past winter, both livestock and pets. Goats and goat kids, cats and kittens, our drake Quack, Mr. Peabody our peacock, and most of our laying hens; it was a slow-motion massacre. One by one, our animals disappeared. We tried everything we could to protect them by securing fences, the barn, and the coop, but it wasn't enough. The coyotes prowled around the property daily, morning, afternoon, and evening. They acted as if they owned the place. It should also be noted that where we live in Canada, it seems that wildlife has more value to the governing bodies than do livestock, and doing away with them is not an option without jumping though a myriad of hoops, something we simply weren't able to accomplish within the small window of time we had.

We stopped letting our farm dog Daisy run loose, just in case

Winter Solace

Taco, mid-shake, getting the snow off his head and body.

coyotes lurked in the bushes where she runs. She pulled against the leash each time one of the sprouts took her outside, unimpressed with her newly imposed restrictions. From the window where I watched, it appeared that Daisy was the one walking our sprouts as they ran after her, tightly gripping the leash. The bright spot in all of this was that OJ, much to the delight of the sprouts and with Mr. Green Thumb's approval, became an indoor cat.

I made the heartbreaking decision to sell our remaining goats so that they would not become a coyote feast. We said goodbye to our two foundation does, Poppy and Clover, and their kids; and we sent away Hawthorn and our senior herdsire, Rowan; sadly, our wethers, Acorn and Gingko, were among the goats that didn't make it. It is painful, not knowing if we will have goats again one day. It is bittersweet, knowing that Poppy and Clover are due to kid again this spring, and I won't be there to watch and support them.

If I'm completely honest, these are the days I feel like throwing in the towel. I hate losing animals, and I wonder what I could have done differently to prevent their deaths. These are the days that I ask myself the same question I hear from well-meaning friends and family: Why? Why do I keep raising animals when it's so tiring, so time consuming, and sometimes so full of hurt and grief? Why do I continue to produce my own food when a twenty minute drive in any direction from our homestead will lead to farmers markets, grocery stores, and big box chains? Why?

During this part of the year, it's not just family and friends asking why. It's also the sprouts as they fight our requests to bundle up and go outside to help with the chores. Why do we have to do it? They complain. We didn't choose to have animals! Why aren't you the one doing it? And silently I answer that I understand. I am second guessing myself too. Before we lost any animals to coyotes, Mr. Green Thumb and I had talked about the possibility of downsizing both our flock of chickens and our herd of goats. The question why echoes in the back of my mind now especially.

I've got easy answers to this question that I throw around when I don't have the time or energy to give a more in depth answer. "I love it," I say. Or, "it's a great way to raise a family." And these answers are true, of course. But they aren't the whole picture. It's hard to describe the deep respect I have for dirt, and seeds, and manure, and compost, and the swelling pride I feel when I bite into the first sun-warmed, homegrown tomato of the season.

Aussie, taking a rest after sledding down our driveway.

Unless a person has raised her own animals, and has loved them deeply, it's hard for them to understand the attachment and connection I feel toward our animals and the joy it gives me to see them grow and thrive and produce food for our family and others. So whenever the question, why, comes up, and we start to seriously consider adjusting this busy, hands-on lifestyle we have created for ourselves, I am reminded that the positives far outweigh the negatives, and I am so thankful for the life we live.

Winter is a quieter season on the farm, which makes it a great time for reflecting on the past year and planning for the year to come. My emotions and doubts set aside, the to-do lists for January and February are written out, and I am working on writing March's list. As I check off each item, I feel we are one step closer to Spring. There are seeds to be ordered and started. There are garden beds to be tilled and animal structures to be erected. There is manure to be spread and piglets to be procured.

These short, dark days are perfect for rereading my homesteading notebook—a haphazard collection of my thoughts about what worked and what didn't over the last year. I've scribbled down notes about our animals, the garden, and the canned, dried, and frozen foods we produced in the hopes that this information will provide us insight into what was successful and what needs to be done differently in the future.

Our freezer is still full of meat—both pork cuts and whole chickens, even after selling two whole hogs. That's good too, because if everything goes as planned, we won't be taking our soon-to-arrive pigs to butcher until August of this year. Due to improper planning on my part, I didn't plant any winter squash and planted no more than a dozen potatoes last summer, so we've had to buy both of those this winter; I'll definitely be giving them prominent places in my garden this coming spring.

We went through the blueberries faster than last year, and I've made a note of that in my journal. Due to an unpredictably dry summer, our blueberry bushes didn't produce as much as we expected, and I had to buy blueberries from a farm across town. This Spring we are planning to install irrigation for both our blueberries as well as our rhubarb in the hopes of producing more fruit. On a positive note, our pantry is still full of jam, as the heat of the summer produced an abundant crop of blackberries—this time last year, our jam was almost gone.

And just when the bone-chilling cold feels like it will never end, seed catalogues start arriving in the mail, and I spend many an evening snuggled up next to Mr. Green Thumb on the couch, perusing their colorful pages and circling all the seeds I plan to order with a bold black marker; these pages are a reminder that as each day grows longer, Spring is truly just around the corner.

Lard

"Everything on the hog is good!"

– Anonymous

For most people, lard is seen as something our grandparents or great-grandparents cooked with; not something we would use in a modern kitchen. Animal fat has gotten a bad rap as a saturated fat that causes high cholesterol and heart disease. However, recently, the tide has been changing, and more people are reconsidering pasture-raised animal fats as a viable, healthy option. For our family and many other small scale farmers like us, using lard in the kitchen is a regular occurrence—part of what we would call nose-to-tail eating; it is simply utilizing all parts of an animal so as not to be wasteful.

Now, I'll be the first to admit that nose-to-tail eating is a challenge for both my husband and me. It's not like recipe books are exactly overflowing with ways to cook tripe, heart, or liver.

My hubby was exposed to a nose-to-tail approach when he was a young child. His grandma was a master chef when it came to preparing head cheese, and my hubby fondly remembers chewing on a chicken foot straight out of a pot of chicken noodle soup. But this type of cooking is a tradition that ended with our grandparents' generation. With the development of massive farming operations and the abundance of cheap meat, it didn't seem worth the effort to cook these parts of the animal anymore.

Raising our own animals for meat has flipped the logic of cheap meat on its head. The first time we raised an animal from the time it was a baby until the day it was butchered, we realized the abundant effort and time it takes to care for an animal that will ultimately feed our family. The strength of my biceps is testament to the number of five gallon buckets of water I hauled (and still haul) daily to quench our pigs' thirst while it was too cold outside to use the outside faucets and hoses. From that first experience, we also recognized how valuable that animal's life is, and the sacrifice it made for the nourishment of our family. Out of respect for that animal's life and sacrifice, our goal is to use as much of the animal as possible.

The first time I rendered lard, I couldn't believe the stench it caused. The smell permeated throughout the house. For days, I could smell the lingering scent of heated pork fat in almost every room of the house. Like the clothes you wear when sitting around a campfire, any clothes that were in close proximity to that bubbling vat needed to be washed, sometimes twice, to get rid of the smell. I learned later that the smell was due to the fact that it had been overheated during cooking. However, at the time, all I knew was that I would smell like bacon grease to everyone I came into contact with for much longer than I would have liked.

Once that first batch of fat had all been melted and strained, I poured the viscous, golden liquid into mason jars and lined them on the counter to cool before refrigerating them. Slowly the warm, translucent fat became lighter and opaque, until it was creamy white in color, and completely solidified. I felt a huge sense of relief at having rendered that large amount of pork fat that had taken up so much space in my freezer. It was another item I could cross off of my to-do list. And yet, at the time, I had no idea how to utilize the rendered fat. For months those jars of lard sat in the back of my second fridge, untouched. They were all but forgotten, buried by various jars of fermented vegetables, cartons of eggs, and other refrigerated pantry staples.

And then one day, someone mentioned making deep fried New Year's cookies, called Portzelky in my family—a once a year donut-like treat my Mennonite grandma used to make when I was a child. I knew immediately what that lard was destined for. I pulled out my almost forty year old copy of The Mennonite Treasury of Recipes, which includes tried and true recipes from at least the turn of the twentieth century—the cookbook I inherited from my paternal grandmother. There were easily half a dozen Portzelky recipes to choose from. I quickly scanned the instructions, and settled on the one recipe that gave actual measurements. The other recipes gave instructions such as, use enough flour to make a soft dough, which to my modern ears and untrained baker's hands, doesn't mean much.

Once the lard was soft enough to ladle, I emptied a quart jar of the cream colored fat into a pot to melt, and then proceeded to make the dough. It was a shaggy, sloppy mess by the time I had completed it, and I didn't know how it could possibly hold together to become a donut in the hot fat, but I was determined to try. I had been preaching to my kids about the humble New Year's cookie as I assembled the dough, and I knew that even if these Portzelky didn't turn out perfectly, I owed it to my kids to finish the project and let them sample the outcome.

The Portzelky most likely did not turn out like my grandma's—although it's been too long since I last tasted one to know for sure; regardless, to me they were sublime. They were the perfect balance of cakey insides and a crisp exterior. Plumped up raisins burst in my mouth with each bite, and the cane sugar coating added the perfect hint of sweetness. The melt in your mouth feel that the lard gave to the little donut-type balls had me swooning.

The kids however, were less impressed. Having never tasted a homemade donut before, and only having the syrupy-sweet Tim Hortons' Tim Bits as a comparison, the flavors fell flat on their unrefined donut palates. I knew then that things would have to change. Homemade treats like Portzelky and fritters would be happening much more often now that I knew how delicious they could be when cooked with love in a pot of humble lard.

Although I was ignorant of it the first time we butchered our pigs, lard can be divided into four categories. The fat from around the kidneys, called leaf lard; the fat from the cured parts of the pig,

A plate of portzelky fritters dusted in sugar and fried in lard from our own pigs.

most notably the bacon, which most people would call bacon fat; the caul fat, which is a lacey fat web that is wrapped around the pig's organs, and the rest of the fat from the pig, sometimes called back fat, but which is most commonly referred to as lard.

The different types of lard have a unique purpose around our homestead. I render the leaf lard and save it for baking delicate treats around Christmas time and other special occasions. The leaf lard is the lightest colored and most mild tasting of the lards. It was only after a customer asked me if I had any leaf lard for sale, that I learned that there was such a thing. A quick google search led me to all sorts of information and inspiration; the next time our pigs were butchered, I asked for the leaf lard to be kept separate, and rendered it extra carefully so as not to scald or brown it.

Bacon fat is perhaps the type of lard used most commonly around our house. I love the subtly smoky depth of flavor it lends to scrambled eggs, fried potatoes, and roasted root vegetables. Not to mention its high smoke point, meaning it can be heated to high temperatures without burning, which is great when aiming for a roasted potato which has a crisp golden crunch. Whenever I cook up some of our home-grown bacon, I strain off the fat into a pint jar I keep on the counter, always at the ready.

Unfortunately my son once mistook my jar of cream colored bacon fat for creamed clover honey, and helped himself to a large spoonful. The awful sound of his retching had me running to see what was wrong. When I discovered the mistake he had made, I had to keep myself from giggling as I helped him swab the fat from out of his mouth and gave him a glass of juice to help rinse the last of it away. Thankfully that experience hasn't ruined his enjoyment of bacon fat infused scrambled eggs—or creamed clover honey, for that matter.

I don't use the caul or back fat regularly, but that doesn't seem to matter, considering once it is rendered, it can keep almost indefinitely in the back of the fridge. I typically use it once I've used up the bacon fat, or to deep fry a batch of potato or vegetable chips, fritters, or battered fish. After learning from that first stinky attempt at rendering lard that overcooking had created the strong, permeating smell I couldn't get rid of, I am much more careful now to check on it to ensure that it doesn't get burnt.

Portzelky—New Year's Cookies

(Recipe adapted from The Mennonite Treasury of Recipes)

Ingredients:

1-1 ½ quarts rendered lard

3 cups white flour

1 cup golden raisins

1 cup water, divided

3 eggs, separated

¾-1 cup sugar, divided

½ cup milk

1 ½ teaspoons salt

1 ¼ teaspoons instant yeast

Instructions:

• Stir 1 teaspoon of sugar into a ½ cup of lukewarm water, and then sprinkle 1 ¼ teaspoons of instant yeast on top of the water, and allow to sit.

• Combine the other ½ cup of the water with a ½ cup of milk, 1 ½ teaspoons salt, 1 Tablespoon sugar, three egg yolks, and raisins.

• Whip the remaining egg whites until they form soft peaks, and set aside.

- Add the yeast mixture to the milk mixture and stir to combine.

- Stir in 3 cups of flour until combined, and then fold in the whipped egg whites.

- Allow to sit in a warm place for a half hour to an hour, to allow dough to rise.

- Pour the remaining amount of sugar into a bowl and set aside.

- Heat the 1-1 ½ quarts in a medium sized pot.

- Test the fat by dropping in a small amount of batter. If the batter rises and sizzles, it is ready. Or, test it with a thermometer; it should be around 375F.

- Drop teaspoon sized amounts of dough carefully into the hot fat, four per batch. Do not use more dough than a large tea spoon scoop, as the dough will expand, and will not properly cook through if it is too large.

- Allow the dough to cook for approximately 2 minutes per side—4 minutes in total.

- Remove Portzelky from the fat and immediately dunk into the sugar, to give it a nice sugary coating and crunch.

- Continue frying batches until all of the dough is used up.

- Serve and enjoy!

- Allow fat to cool slightly, and then strain it into a quart jar. Store it in the fridge and reuse it several more times for frying if desired.

Homemade Herbal Salve

"[H]erbs truly shine as a home health care system. When you study and connect with the plants, you can rely on them, and yourself, for most of the ailments that you and your family will experience."

- Rosemary Gladstar

As our sprouts spend time exploring blackberry laden valleys, and practice their fire-building skills; as they help repair field fencing alongside Mr. Green Thumb, and attempt at cuddling the semi-feral barn cats, bumps and scrapes, bruises and burns are bound to happen. Some days they come to me, asking for a salve to put on their ouchies. Other times, they will complain of their injury, only to run and hide when I suggest making a comfrey and plantain poultice for their wounds, or threaten to wash their sores off with chamomile and yarrow tea.

Although some of these concoctions may sting a little, which is why my sprouts sometimes run away at the suggestion, these herbal remedies are mild and gentle medicinals. They offer natural alternatives to many over the counter rubs and creams. With a little forethought and preparation, these herbal remedies can fill the medicine cabinet and be available at the ready to provide healing support to all types of minor injuries and ailments.

One of the best ways to capture the essence of these herbal medicinals is to make an oil infusion with them. Dried plant material is best as the water in fresh leaves, flowers, or roots can cause the oil infusion to mold. I usually harvest fresh plants from the garden or around our homestead and then bring them in to dry. If it is the aerial parts of the plants I harvested, I give them a quick rinse and allow them to air dry in an out of the way place, and out of direct sunlight. If it is the root of the plant, I wash it more thoroughly, cut it into roughly half inch pieces, and dry them at the lowest setting of my oven until they have shriveled and are firm to the touch.

From these infusions, myriad of products can be made. In soap recipes I can replace the regular oil with an infused oil for additional cleansing properties without having to alter the recipe; lotions and body butters can incorporate these medicinal oils to create deeply nourishing products for daily use. Lip balms, facial cleansers, and hair treatments can all benefit from the boost in nutrients from infused oils. My favorite use for these oils though, is in salves.

By preparing homemade salves with infused oils, I create customized formulas to treat each specific ailment. For example, Arnica should only be used on skin that does not have an open wound, but it does wonders in reducing the pain and duration of bruises. Comfrey root is similar in that it should only be used on unbroken skin, but aids in the mending of fractures and broken bones, as well as less traumatic injuries such as bruises or bumps. Calendula supports all types of skin conditions as well as injuries and is found in most of my formulations; Yarrow is fabulous to

Homade Herbal Salve

Making soap with my sister-in-law, Becky. *Photo by Becky Billion.*

help stop bleeding, and was historically known as soldier's woundwort; it also has properties like Calendula to aid in skin treatments. Chamomile and Dandelion flowers help reduce pain, while Lavender has antibacterial properties.

When combined, these herb-infused oils create the base of the salves I make, which provide the body with the support it needs to promote healing. While all of these specialized salves are handy to have at my fingertips, I tend to use my All-Purpose Herbal Salve the most. I use it on the rough, cracked skin on my feet all summer long, and carry it with me everywhere I go to treat any unexpected ouchie that happens along the way. What I love most about this salve is that all of the ingredients are so mild that children as well as adults are free to use it daily. A friend of mine used several jars to help clear up her son's facial eczema. I've even used it as a lip balm in a pinch.

Oil infusions can be made three different ways: cold infusion, warm infusion, or double infusion. Cold oil infusions are created without the use of heat by combining the plant material and oil into a jar, and allowing time to enable the oil to extract all of the plant's fat-soluble nutrients. The jar is capped and set in a cool, dark place for four to six weeks during this process. A warm infusion enlists the help of a temperature increase, as opposed to time, to extract the same nutrients. Typically this is done in a double boiler, which can be as simple as setting the jar full of plant material and oil, uncapped, into a pan of gently simmering water for roughly an hour.

The double infusion can be made by employing either of the above methods, but doubly concentrating the oil by straining out the original plant material, and then placing more plant material into the oil to infuse it again. Instead of precise measurements, I tend to use the folk method, a fanciful way of saying that I eyeball it. The folk method for making oil infusions is to simply place plant material into a vessel, and then cover it with oil. I haven't measured to be sure, but it seems to work out to about a 1:2 ratio. Any natural oil could work for this, although I tend to use olive, coconut, and avocado oils for their own skin nourishing properties.

All-Purpose Herbal Salve

For the infused oil:

2 ¼ cups olive oil, coconut oil, or avocado oil (or a combination of the three)

- Prepare a single or double infused oil with dried dandelion petals, plantain leaves, calendula blossoms and lavender blossoms, as described above.

For the salve:

½ cup beeswax pastilles or grated, packed beeswax.

- Combine 2 cups of infused oil with ½ cup beeswax in a glass jar or measuring cup that you don't mind having a waxy residue on afterward.

- Place the glass vessel into a pan of simmering water and heat over medium low heat until beeswax begins to melt; turn heat to low and stir with a stainless steel utensil until all the wax has melted.

- Take the glass vessel out of the simmering water and immediately pour into five half cup (125 ml) mason jars, scraping the sides as necessary, as the mixture will already begin to harden; allow to cool completely before capping.

- Use on any scrapes, cuts, bumps, bruises, burns, bites, stings, sunburns, eczema, dry skin, etc.; safe for use daily; safe for use on children; stop use if a rash or irritation develops.

Chicken Dinner

"Winter blues are cured every time with a potato gratin paired with a roast chicken."

- Alex Guarnaschelli

I didn't learn to cook growing up; I learned to cook while on sick leave during my first pregnancy. I watched the Food Network as if it were my full time job, and I spent a good deal of time experimenting in the kitchen whenever my stomach could handle it. My eyes were glued to the screen as chefs cut apart chicken carcasses, baked chicken breasts with lemon and fresh herbs, prepared all sorts of dishes such as pulled or jerk chicken to white chicken chili. I even took notes about each chef's tips and tricks. But the piece-de-resistance was the perfectly prepared, golden roasted chicken with gravy.

However, I never attempted to roast a chicken myself in the early years. It felt unattainable. Partly because the idea of bringing home a raw, cellophane wrapped, grocery store chicken didn't appeal to me; but also because I didn't feel grown up enough to pull off such a meal. It seemed to me that this was a dish best prepared by the matriarch of the family for a Sunday afternoon feast set with silverware —not something I threw into the oven from frozen with a screaming baby on my hip. I wish I had known then just how easy it actually is.

When we moved to our homestead, I was intent on learning how to prepare a roast chicken. It seemed like one of those homesteady things I should know how to do. Since we had not butchered any of our own chickens yet, I set out to source free range chicken from a local farm and I was happy to find one within a fifteen minute drive from our homestead. I left their farm store with two frozen shrink wrapped chickens; one destined for the oven, and one as backup.

I pulled out my already worn copy of the *Joy of Cooking*, which I had received as a much appreciated gift several years earlier. I wanted this roast chicken to be perfect. As it sat soaking in a bowl of warm water in the sink, I set to preparing the bed of carrots, potatoes, garlic and onions that it would rest on. Once that was ready, I placed the chicken carcass on top of the vegetables in the roasting pan, and gave it a sprinkle of salt, pepper, and dried thyme, and shoved the tray into the 375F oven. By the time Mr. Green Thumb arrived home from work, the aroma of the roasting bird was wafting through the house, although I still wasn't sure what the final outcome would be.

He and I both stood expectantly by the oven door as I opened it and pulled out a golden brown roast chicken. I did it, I thought to myself. We hovered over the cutting board where the chicken was resting, carving knife in hand, anxiously waiting for the juices to settle back into the meat. We nibbled at an au jus soaked carrot, and picked at the crispy salted skin, and finally, I did the honors of plunging the knife deep into the carcass to begin the process of breaking down the bird, as I had watched so many Food Network chefs do before me. The knife got about half way in, and then it stopped. It wouldn't budge. It was still frozen solid in the center.

One of our Australorp hens, curiously exploring in the snow.

I let out a deep, exacerbated sigh, and placed the icy bird back into the roasting pan, and into the oven.

    I don't remember exactly what we ate for dinner that evening, but it certainly wasn't roast chicken. It took about another two hours before it was cooked completely through, and my four little sprouts would not have had the patience to wait that long for a fancy chicken dinner when they would be just as satisfied with pancakes or grilled cheese sandwiches. I do however, remember eating the roast chicken straight out of the pan later that night. We lingered in the silence of our kitchen after the kiddos had been tucked into bed, and ate at least half the roast bird before stopping ourselves and putting it away for the following night's meal. It was that good.

The second memorable roast chicken dinner was the first time we cooked our own bird. I prepared it much the same way I did the first, a splash of olive oil, salt, pepper, and some thyme and rosemary, fresh from the garden, minced up and sprinkled on top. By this time, I was a pro at cooking whole poultry, although cooking a young, succulent meat bird is nothing like cooking a tough, old hen. From what I had read on several homesteading blogs, low and slow was the method of choice for roasting our homegrown hen.

It smelled heavenly as I pulled it out of the oven; it looked perfectly browned, and its juices swirled around in the bottom of the pan as I carried it to the countertop to let the carcass cool, and to prepare the gravy. The skin was delightfully savory and crispy, and I helped myself to several taste tests while whisking the simmering au jus on the stovetop, before taking it off the heat and stirring in a cornstarch slurry. I expertly removed the leg and breast meat, artfully arranged the pieces of meat on a platter, and poured a generous ladle full of gravy over top. It felt as though our family was dining like royalty, and I was bursting with pride at the thought of putting our home-raised bird down in the middle of the table.

The sprouts oooh-ed and ahhhh-ed convincingly as I placed a serving of both white and dark meat on each of their plates. And then the moment of truth arrived. We began taking our first bites of this special bird, and savoring the umami of the meat as well as the moment. But as we all began to chew the meat, we noticed something unappetizing about the texture of the bird—as much as we chewed and chewed, none of us seemed to be able to swallow the dense, stringy, toothsome meat. This was certainly not how I

Chicken Dinner

Roast chicken dinner, a regular meal in the winter months on our homestead. It can be stretched to make many meals, including hearty soups.

had anticipated this moment. Magical? Yes. But disappointed? That was not a feeling I had anticipated. How were we supposed to butcher and eat our own birds if they were as unappealing as this?

I was determined to find a better way to prepare this precious meat so that we would be able to both nourish our bodies and enjoy the process of doing so. As I was recounting this story to a family friend, she referred to our old birds as stewing hens, and I immediately knew how I needed to prepare these birds in order to use them best—they weren't meant for the roasting pan; they were meant for the soup pot. Low and slow took on an entirely different meaning as the chicken carcass, feet and all, simmered in an immense stock pot on our gas stove, filled with all sorts of vegetables and aromatics.

Truly, that broth was unlike any other chicken broth I had ever tasted. The rich, succulent, complex flavor of the bone broth that we made that day, and continue to make with our stewing hens, never grows old on me. At one point, it seemed inconceivable that something so aged and spent could produce such an outstanding flavor. Although I am exceedingly grateful that such a simple process could produce a meal as humble and yet exquisite as bone broth.

Roast Chicken, Gravy and Broth

Ingredients:

3-4 pound whole chicken (fryer/ roaster)

1-1 ½ Tablespoons olive oil

2 teaspoons dried oregano

2 teaspoons dried thyme

1 teaspoon dried sage

2 teaspoons salt

1 teaspoon pepper

Instructions:

- Preheat the oven to 350F.

- Place the unfrozen chicken, breast side up (drumsticks down), into a roasting pan and drizzle with 1-1 ½ Tablespoons olive oil, and sprinkle with herbs, salt and pepper.

- Place the pan into the middle rack of the oven, uncovered.

- Bake a 3 lb. chicken for 2 hours; add approximately 40 minutes for each pound the bird weighs—if it is

4 pounds, roast it for 2 hours and 40 minutes.

• To be sure the chicken is done, use a thermometer to check that the internal temperature of the meat is 180F.

• If wanting a really crispy skin, turn the oven to broil for several minutes, keeping a close watch on it.

• Allow the chicken to rest for 15 minutes before carving.

• To make gravy, strain the liquid from the roasting pan into a saucepan and bring to a boil.

• In a small bowl, mix 1 Tablespoon of cornstarch with 2 tablespoons of water making sure there are no lumps.

• Pour the cornstarch slurry into the chicken liquid, and stir for several minutes while boiling, until the gravy thickens. If the gravy is too thick (the bird didn't release much liquid) add some water to the gravy to thin it out to a better consistency. If you prefer a thicker gravy, add another Tablespoon of cornstarch mixed with water to the gravy. Taste the gravy, and add salt and pepper if necessary before serving.

• To make chicken broth, place the bones and the liquid (or leftover gravy) into a stock pot or slow cooker, and add 8-12 cups of water. Simmer on low for several hours up to all day on the stove, or overnight in a slow cooker. Strain out the bones, and use as the base of a soup.

Winter Elixir

"Let food be thy medicine and medicine be thy food."

– Hippocrates

Winter is the most common time for colds and flus, and although our family members have strong immune systems from healthy eating and plenty of exercise, time spent outdoors, and being around animals, we are not completely immune. We have been hit particularly hard this past year with numerous viral and bacterial illnesses. Sore throats seem ongoing, and the coughing has dragged on for weeks. There's nothing more unpleasant than bundling up and heading outside to do chores while sick. The chill from the biting wind accentuates the aches and pains. The kids and I spend time in front of the fireplace warming up after chores at the best of times, and linger extra long if we aren't feeling well. I've often just gone back to bed after chores while sick, in the hopes that I can sleep most of it away.

Before turning to over-the-counter solutions, our family focuses on nourishing foods like homemade chicken soup, healing and soothing herbal teas with echinacea and licorice, and homemade herbal syrups to help our bodies heal. Our first go-to remedy for a scratchy throat is a teaspoon of raw honey. Its antibacterial and anti-inflammatory properties help relieve most throat irritations, and it goes down easily for kids and adults alike. Cocoa and cinnamon are also said to help ease sore throats, and occasionally I'll stir a little of one into the honey I give to my sprouts for a treat.

I love the idea of being able to use food not only to nourish but to help heal. Common kitchen herbs such as thyme and sage, for example, help relieve a cough; ginger is an immune booster and helps relieve nausea. Mint and fennel can ease stomach pains; and garlic is antibacterial and antiviral, and can help wipe out a cold or the flu. While I swallow chunks of raw garlic when I am sick, my kiddos cannot stomach the notion. They often end up with slices of garlic inside their socks before bed instead, so that their bodies can absorb all of the beneficial nutrients and bacteria and virus fighting properties through the skin of their feet and into their bloodstream.

Rose hips are another example of a common plant providing nourishing medicine. All roses, both cultivated and wild, produce rose hips, the fruit of the plant. The red or orange globes that cling to its branches all winter long are full of seeds, ready to multiply when given the chance. The globes, or rose hips, are also full of vitamin C, and have been used for centuries to prevent scurvy throughout the winter months. Steeped whole in hot water is the easiest way to access the rose hips' nutrients, although I could also take the time to halve each hip and scrape out the hairy seeds

Rosehip syrup, high in vitamin C, made by boiling rosehips in water, and then adding honey to sweeten and preserve it. This will last in the fridge all winter.

found inside. The halves could then be dried, or turned into jam. Only a little sugar needs to be added to the cleaned halves, and hot water to cover, and after sitting on the counter for several hours, the fruit will have set in the jar—no need for pectin or citric acid as the fruit is naturally high in both. I collect the hips each year from mine and my family members' cultivated rose bushes, and I have a jar full in my apothecary cupboard to use in my winter elixirs.

Another simple, delicious way to help boost our family's immune system each fall and winter is to take elderberry syrup. Said to be one of Hippocrates favorite herbs, elderberry is used to treat numerous ailments in herbal medicine, but it is most commonly known for its cold and flu fighting properties. Both the flowers and the berries are beneficial to make the syrup, but I find the dried berries easier to buy locally than the dried flowers, so I usually make my syrup with just the berries. I have also seen many wild, red elderberry bushes along roadsides and train tracks as I drive, but I have yet to spot a location where it would be simple to pull over and harvest from them.

I had planted five little elderberry bushes next to our blueberry patch one year after we moved to our homestead. Unfortunately they looked more like six inch tall, stubby sticks in the berry patch the following spring and were mowed down by an unsuspecting volunteer, as they were too small to be noticed among the tall grass. (I also lost sixty strawberry plants, and about a dozen young currant and haskap bushes on that fateful day, but who's keeping track?) I replanted the elderberries again this past year, with extra tall stakes, in the hopes that they will actually grow to maturity this time around. I am looking forward to having jars of homegrown elderflowers and elderberries in my apothecary cupboard.

Winter Elixir: Elderberry Syrup

Ingredients:

¾ cup dried black elderberries, or 1 ½ cups fresh, stems removed

½ cup dried elderflowers, or one cup fresh, optional

½ cup dried rose hips or rose hip pieces, or 1 cup fresh rose hips, optional

3 cups water

1 cup unpasteurized honey

½ cup raw apple cider vinegar

Directions:

• Combine berries with flowers and rose hips if using, into a pot along with three cups of water. Bring the water to a boil, and then simmer uncovered on the lowest setting for half an hour.

• Strain the mixture, pressing out as much liquid as possible, and pour liquid into a 1 quart jar. Once the liquid has cooled to room temperature, add one cup honey and stir until combined.

• Add the raw apple cider vinegar.

• Keep refrigerated, and shake before use.
• Take one teaspoon daily, or one teaspoon three or four times daily when sick.

Food Is Love

"Food is not just fuel. Food is about family, food is about community, food is about identity, And we nourish all those things when we eat well."

- Michael Pollan

Winding down the tree lined driveway to the property that would become our homestead, my breath caught in my throat. Mr. Green Thumb and I glanced at each other wide-eyed as the driveway opened up to reveal the house that would become our home —a smokey grey Tudor with white trim, which was built almost fifty years earlier. Butterflies frolicked inside my stomach as we rapped the knocker. Antique double doors ushered us in, and the smell of garlic and onions rose to greet us from the kitchen. The matriarch of the home led us into her living space and offered us freshly baked cookies and tea. Her husband and son, who were hovering over an enormous stainless steel pot bubbling on the stove, walked around the counter to warmly welcome us. As we sat around the table at our host's request, she said something that has stayed with me to this day: In our family, food is love.

This was obvious that day in her family as they cooked and shared their abundance with us, their guests; I believe that it is evident today in our family as well. One of the primary ways my husband and I strive to show our love to all of our guests is to intentionally grow, passionately cook, thoughtfully eat, and generously share the abundance of the harvest with those who come into our home. There is nothing quite like preparing and sharing food with those you love. Since living on our homestead, we have always had as our goal to welcome family and friends, new and old, into our home and to our table in the hopes that they will feel loved as we eat together.

In the early days, I didn't always do this well. My house may have been spotless, and the food on the table a lavish spread of dishes created from homegrown food, but on the inside, I was a ball of nerves. I was often so stressed that I could not enjoy myself or the company of our guests; I wanted everyone else to enjoy themselves at my expense. These days, our hosting looks a little different. I'm not concerned about a few dishes in the sink, or if everything comes out of the oven at exactly the same time. We often have a simple appetizer sitting on the counter while we wait for our food to finish cooking; I usually enlist one or more of our guests to help prepare the salad, or even harvest a few veggies from the garden for one of the sides.

Mr. Green Thumb and I take a much more relaxed approach to hospitality now. It's not that we don't care about those who share our table, or what we serve; we have simply learned that when we are relaxed, our guests will most likely sense that, and relax themselves. Everyone enjoys the time together when their hosts are relaxed. I still don't do this perfectly. When those with

Rae-Rae cradles a gigantic head of Red Russian garlic. Each garlic clove will be planted in the autumn in order to produce a new head of garlic the following summer.

A modest harvest of quince, from our young tree.

food sensitivities or food preferences which are different than my own come to dine, I often feel at a loss; but I am humbly reminded that sharing our home-raised meat or farm fresh eggs, as fun as it is, must be less important than those at our table. Extending our table to those different from us is one of the best ways to find common ground, to connect in a way that builds bridges and deepens friendships.

Mr. Green Thumb and I have also pushed ourselves beyond our comfortable limits by inviting new friends over who don't look, speak, or eat like we do.

We have learned a lot about hospitality from our Syrian friends who immigrated to the Fraser Valley just two years ago. Their approach to serving meals to family and friends has been a delightfully refreshing experience. Although there have been times of miscommunication and awkwardness, these moments have stretched and strengthened our friendship with their family. Breaking bread with those who speak, look, and eat differently than we do has helped us to be more thoughtful, considerate hosts to all of our friends and family.

Some of the foods we serve also remind us of our history and of the places our families come from. Our ancestors, like we do, valued the hard work of preparing and bringing in the harvest and of sharing the table with both friends and strangers alike. They were true homesteaders, relying on everything they grew, raised, and produced to get them through each season, to feed and clothe their children, to keep the fuel in their fires going and the stove tops hot and ready to cook. Their hospitality may have been a more simple version of what ours is today, but it was no less authentic. They generously shared what they had, however humble,

My growing apothecary, filled with jars of dried herbs, dried flowers, and infused oils, waiting to be turned into tisanes, salves, balms, etc.

just as we strive to do today.

The bread and buns my grandmas both baked, the soups my mom prepared, the jars of pickles, fruits and jellies that lined my grandma's pantry shelves, the fresh ingredients that were harvested and woven into meals—all of these showed me as both child and adult alike, that I was loved by my family. All of these ingredients have their place in the homestead kitchen and pantry, but they also belong on the homestead table; they weren't made to simply be admired or so that I could feel accomplished; they were made to be enjoyed and savored, satisfying and nourishing—they are a demonstration of love to each one who partakes.

Just as the Creator made plants to grow, and chickens to lay eggs, and weeds to provide nourishing medicine to demonstrate His love for us, we were made to share these gifts with those around us to show love to them. Whether a new or lifelong friend; whether a neighbor down the street or a neighbor whom I sit next to at my sprout's baseball game; whether my own family or the immigrant family on the other side of the city; sharing a meal with those around us, a generous portion of meaningful conversation served up alongside dishes of nourishing food, is one of the best ways to find commonality and friendship; sharing a meal is one of the most inviting ways to extend a welcome to whoever steps over our threshold.

Roasted potatoes from the garden, and our home raised pork with gravy--sharing the abundance of our homestead around our table.

An Ode to the Animals that Left too Soon

"All things bright and beautiful, all creatures great and small, all things wise and wonderful, the Lord God made them all."

-James Harriot

Saying goodbye to the precious creatures we are entrusted with caring for is never easy. For many different reasons, we have had to say goodbye to many animals that lived on our homestead over the last five years. Our sprouts have learned that death is a part of living, and they are learning alongside Mr. Green Thumb and me, how to come to terms with the grief that these animals leave once they are gone. It didn't feel right just to mention these animals' deaths, disappearances, and even their moves for various reasons, without also acknowledging the impact they had on our lives.

2015

Goodbye, Puffy and Fluffy, our two sweet bantam hens—two of our first ever chickens. You lovely ladies hatched out the first batch of chicks on our farm. We still have some of your granddaughters laying eggs for us.

2016

Goodbye to our precious pup, Scarlett. For three long years I watched you grow and mature. You were about the best farm pup a girl could ask for. You were constantly by my side, cheering me on while I completed my chores. I wish I knew why you disappeared. The kids and I like to imagine that a very loving family found and adopted you. You are forever in our hearts.

Goodbye, little duckling. You tugged at Aussie's heartstrings and opened wide his eyes to the tenderness that exists between animals and the humans who stop to notice.

2017

Goodbye, you seven, stupid turkeys. It seemed your only mission was to self-destruct.

Goodbye, Taco Cat. You came to us with a chip on your shoulder, and a tendency to lash out, but once you had learned to trust us, you became a fiercely loyal companion and friend. You didn't catch as many rodents as we would have hoped, but you were a lovely partner to keep me company in the garden and while doing farm chores.

Goodbye, Roxy. I'm sorry that we had to send you away. You were a sweet girl with a big heart. You earnestly tried to be a good little farm dog, but your instincts got the best of you and so you had to find a new home. I'm glad we get to visit you once in a while.

AJ, snuggling up to OJ, our newly appointed indoor cat.

Goodbye, three chickens. I'm sorry Roxy mistook you for chew-toys and unwittingly mauled you.

Goodbye, Cosmo. You were a playful, loving young tom cat. You were equally good at hunting mice and cuddling on my lap while I sat writing this book. You even tried to write a few words of it yourself when you strolled across my keyboard. I cried big, ugly tears when we found your cold, stiff body on the side of the road. You are missed.

Spring 2018

Goodbye, Violet. You were such a gentle creature. Your big blue eyes used to look up at me so trustingly. You would nuzzle your nose into my hand as I gave you ear scratches. I stroked your head and your ears as you drifted peacefully away.

Goodbye, Lavender. I'm sorry I didn't know your mama was sick, and that she passed her illness onto you. You were a gentle and timid creature, much like your mama, from what I can tell in the short time I knew you. I'm glad you didn't have to suffer.

Goodbye, Brownie the second. You were a sweet quadruplet, born into goat-heaven. Your sisters are beautiful does, strong, calm, and trusting. You would have made a handsome herdsire.

Summer 2018

Goodbye, Latte. You were such a special companion to Rae-Rae. Your gentle spirit and calming presence were just what our little sprout needed. You were a loving pet and tender mother who always looked after your babes. And then one day, you joined Mocha and Sugar Cube on a rabbit-hunting expedition and never returned. You are living on in your two beautiful boys, Chino and OJ.

Goodbye, ducklings and chicks. Every day you toddled along behind your mama hen, pecking at the dirt and weeds, and the kiddos and I would count to see how many of you there were. Every day, it seemed, there were fewer of you, until you were all gone. I'm not sure if it was the rats or the cats that got to you.

Goodbye, Timber. You were always exploring, never fearful. I

Mr. Peabody, shaking his teenaged tale feathers for the hens.

had high hopes that you would be a bold huntress like your mama, Mocha. You were such a charming kitten. It's too bad that in this case, curiosity did get the cat.

Goodbye, Lilikoi, our beautiful fluff ball, Aussie's first and last 4-H rabbit. He bought you with his own money and tenderly cared for you daily, until one day we found your little body limp and lifeless. You didn't like to be brushed or have your nails and wool clipped, but you were happy to have ear scratches and cuddles. You loved hopping around on the grass, and nibbling on dandelion flowers and leaves. It's too bad we weren't able to take you to any fairs this summer to show everyone just how perfect a French angora rabbit you were.

Fall 2018

Goodbye, to the three goat kids who had not yet been named. You had families that were eagerly awaiting the day they could bring you home. Why did you think the grass was greener on the other side of the fence? Why did you have to find the one gap between fence and ground that we had failed to notice and secure? Why did you have to act like typical teens who think they're invincible and don't listen to their mamas' warnings? Why?

Goodbye, Mr. Peabody. You were a handsome boy, full of spunk, and yet graceful; assertive and yet gentle. When I saw the two piles of your splendid feathers in the neighbor's field, I broke down. I imagined how hard you must have fought to get away from the coyote, only to have him capture you again. Mr. Peabody, you were a beautiful creature. I loved having you strut around our yard; I loved seeing you proudly perched on top of our chicken coop. I hope one day we will be fortunate enough to have another of your kind living on our homestead.

Goodbye, Sugar Cube and Fluffy. You were a radiant, loveable mother daughter duo. CJ loved you to pieces, and was heartbroken over your disappearance. She has since named two kittens in your honor, and loves them deeply, as she loved you both.

Goodbye, to the forty chickens we rescued this summer. You clueless, wandering birds fed the local coyote population well.

Goodbye, Misto. Mr. Green Thumb was driving really slowly and didn't see you—contrary to what Rae-Rae has argued.

Cosmo, up to his usual playful trouble, temporarily became one of the Christmas ornaments on our tree.

Goodbye, Poppy and Clover, and Poppy's baby which is now named Melanie. Although it was bittersweet to say goodbye, I know you are being well loved on your new farms. I love that the farm girlies and I get to visit you from time to time. I hope you know that we still love you deeply.

Winter 2019

Goodbye, Chino and Mocha. You were both special cats in your own ways. Mocha, you gave us two litters of strong, healthy, brave kittens. Most of them moved on to other families and homesteads, but the three we kept from your litters are fierce and independent, and about to have babies of their own. You would have been such a proud grandma. Chino, you big ball of fluff, oh how I miss our morning snuggles. It didn't matter to me that your long hair got all over my clothes; you knew how to cheer me up on days I didn't feel I could be happy. You are so missed.

Goodbye, Acorn and Gingko. My heart still thumps in my chest when I think about how you both went. Oh, you poor babies, the distress you must have felt as the coyote chased and caught first one and then the other. I had nightmares of you innocent boys being hunted down. I am so sorry I couldn't protect you from that awful fate.

Goodbye, Rowan. My handsome herdsire, I am so, so sorry that you watched with horror as your friends, Gingko and Acorn were attacked by a coyote. You were shivering all over when we found you, hunched up in a corner in the barn. I hope you know we sent you away to protect you. It sounds like you are happy where you are, being pampered like a prince.

Goodbye, Quack. You always made me smile when I saw you waddling around the homestead. I loved how your sleek black feathers shone green when the sun hit them a certain way. You loved and protected your ladies well, my friend. You have four adorable ducklings, and I hope at least one grows up to look just like you.

Spring 2019

Goodbye, honey bees. It was a sad day to discover that your colony had not survived this unpredictably harsh winter. I hope I can prevent that from happening again with what I learned from your hive. I still have several jars of honey in my pantry that you made, and I savor each spoonful with great appreciation and delight.

Goodbye, OJ. Darling OJ. We brought you inside to protect you from the coyotes, and you quickly became friends with us all, even Daisy. You two would play together so well. She looked for you all over the house for days after you disappeared. I really thought you would come back, but it's been many weeks now, and I think you must be gone for good. I'm sorry friend.

There cannot be life without death. It has been a hard lesson to learn for us all. This list grew extensively as I continued to write this book, and reading it all in one sitting makes me feel as though this is the most notable thing that has happened on our homestead—although I know this is not the case. There has also been a lot of living that has taken place, marvelous living. And so, I am reminded of Alfred, Lord Tennyson's words as I reflect on the creatures I memorialized, "Tis better to have loved and lost than never to have loved at all."

Spring Again

"No winter lasts forever; no spring skips its turn."

- Hal Borland

 Spring is the glimmer of hope on the horizon that comes after a long winter's night. Spring is inhaling deeply of rain drenched earth and cherry blossoms. Spring is watching goat kids bouncing in the field, and piglets uprooting their pasture. Spring is the sound of song birds chirping from overhead branches, and a pair of bald eagles returning to nest in the neighbor's tree. Spring is sowing radish, arugula, and pea seeds in the garden and checking on them daily to see if they have sprouted. Spring reveals new life, new growth, renewed and hopeful anticipation. Spring is my favorite season.

 This year especially, I needed the newness of spring as a balm to soothe the sorrowful losses I endured these past few months. This spring has brought new life to our farm, both in expected and unexpected ways. One of our Muscovy hens began sitting on a clutch of eggs about a month after our drake Quack was taken. I didn't think her eggs would still be fertile and was annoyed at her for ending her short laying stint by getting broody. However, roughly thirty days later, she successfully hatched out four fluffy little ducklings.

In past years, neither our chickens nor our duck hens have been able to keep their ducklings and chicks alive without our intervention; I was reminded of this just this past January when she had hatched out five ducklings, which all disappeared within several days of us discovering them. These four fluffy ducklings were the last of Quack's babies, and I didn't want to lose these too. We set up a small pen with the mama and her ducklings, and hoped for the best.

We have also been on baby watch with our pregnant queens. I wasn't sure they had survived the winter or the coyotes, until I spotted each of the three of them nimbly jumping and climbing fence posts and trees with bulging bellies. One of the three pregnant mousers born on our farm last year has all but decided she'd like to be a house pet, and for now, we'll let her. She is as eager for head scratches as the sprouts are to give them to her. Her uncharacteristically friendly nature, coupled with the fact that her teats and belly are plump and round, alerted me to the fact that her time is getting close. And so we wait, and watch.

The demand for our pasture-raised pork has increased each year, and this year is no exception. We bought six heritage-cross bacon seeds from a local farm and moved them to the homestead in late winter; one for us, and five to sell as quarters and sides to both new and return customers, all eager to fill their freezers with delectable, pasture raised pork. This is the most we have raised at one time yet. With each month we have had them, the pigs have grown longer and rounder, and I cannot help but see bacon and hams flash before my eyes whenever I look at them.

After a few months on pasture, the pigs had upturned most of the grass, and so daily, along with food and water, my sprouts

A spring rainbow, following a heavy rain, beyond our garden. It is a reminder for me that good things are ahead, in spite of the challenges we have faced on our homestead.

and I gather a wheelbarrow's worth of handfuls of hay growing in the field, roots and all, to throw to the expanding porkers. The pigs are immensely pleased to find this pile of long grass in their paddock, and they contentedly chomp up the entire heap in one sitting.

We have also decided to experiment with pastured goat and lamb this year. We brought home a pair of each to keep in the barn where our Nigerian Dwarf goats used to be housed. It is bittersweet to hear the goats call out at they stand at the top of the mound of dirt where our brazen buck Rowan used to stride. The goats have begun making a dent in the wall of blackberry bushes hanging over the east facing fence, while the lambs are content to munch from the selection of grasses growing up from the pasture. We will keep them a few months longer than the pigs. Their diet, made up almost exclusively of grass and forage, doesn't allow these animals to grow as quickly as if they were fed a steady diet of high carb fodder and grains. By late fall they will be ready to head to the butcher. On a whim, we picked up a handful of guinea keets when we picked up some chicks to increase our diminished numbers. I have yet to see adult guinea fowl in real life, although I have oohed and awed over other homesteaders' facebook photos and posts about them. All I know is that they are severely noisy, immensely hardy, eager foragers, and consistently reliable to come back to roost. In theory, these birds sound like a perfect fit for our homestead.

Our girlies Rae-Rae and CJ came with Mr. Green Thumb and me to the local poultry swap where we picked out our birds. They were enamored with all of the tables stacked high with egg cartons of hatching eggs, crates full of chicks, poults, keets and ducklings, and the occasional fully grown bird. They have decided to start breeding poultry and to start setting up their own booth at the poultry swap this fall. If our guinea fowl make it to adulthood, they will be able to breed these, along with the chickens, ducks, and geese that currently reside on our homestead.

The bees have settled into their home and are unassumingly gathering pollen and nectar from every flower that is in bloom on our homestead and around the neighborhood. The kale and mustard greens that have overwintered are sending up flowering

Aussie, CJ, AJ, and Rae-Rae, posing in front of our newly planted garden beds, just before they begin their Easter egg hunt.

stalks in preparation to produce seeds, and they are covered in honey bees and wild bees alike. The pollen baskets that honey bees naturally carry on their thighs are full of bright orange and yellow pollen as they land on the front doorstep of their hive; the frames in their box are beginning to fill with honey alongside the comb filled with larvae and pollen. Soon, it will be time for the honey flow, and I will be adding a honey box on top of their current box so that they have enough space to store all of the honey they produce.

Our fruit trees, which are in their fourth spring in our small orchard, are all bursting with green growth. Their swollen buds have given way to unfurling green leaves; their delicate, white flowers have blossomed and are now dying back to reveal small, hard, green, pebble-sized fruits. While all of the trees have set fruit, there are several trees covered in tiny fruits, including one of our hearty figs, and we are anticipating the feast we will enjoy when they are ripe.

We have added several Asian fruit varietals to our orchard over the last couple of years as well; a Yuzu citrus, two Asian pears, and Fuyu persimmon. They are growing well, and to our excitement, and with a bit of extra protection, have survived the last couple of winters. Our persimmon is especially thriving; it is covered in tender, glossy, chartreuse leaves. We are hopeful that next spring we will see flowers and fruit alongside the leaves on this exquisite tree.

This spring, as every season does, has brought change to our homestead. A harsh winter, both physically and emotionally, has caused Mr. Green Thumb and me to think deeply about the lifestyle we have chosen, and to reflect on how far we have come both in terms of knowledge and skills, and the life we have made for ourselves and our sprouts. We have long pondered the direction our homestead will head in the future—whether or not we will return to goat breeding again next spring; whether we will continue to sell eggs to the public (since losing chickens seems to happen far too easily); and whether we will add cut flowers to the list of products we sell.

Homesteading is not for the faint of heart, although I have felt faint of heart often since homesteading. It is physically, mentally, and emotionally taxing. Although falling into bed after a long day of outdoor laboring or indoor jam making has its appeal, to do this daily requires more than an acreage and a yearning for the country life. It demands equal parts grit and romance—seeing through both the sweat of my brow and through rose-colored glasses. I am still trying to decide if I have what it takes. In the meantime, the animals need food, the garden needs weeding, and the food in the freezer needs cooking, and so I get up, and I live this homesteading life we have set out to enjoy.

I was recently talking to a friend about our lifestyle, and how after five years of homesteading, I feel as though I am legitimately a farmer now. CJ was playing within earshot and piped up, "Mom, you're not a real farmer. I've been doing it my whole life—I'm the real farmer."

Yes, baby girl, you are the real farmer. This is the only life you have known, and you are living it well. And really, that was always the point of this homesteading journey.

It wasn't only to live this lifestyle, although that is rewarding in and of itself; it was so that Mr. Green Thumb, our sprouts and I would all learn to value more deeply the land around us, the food on our plates, the joy of a hard task well done, and the abundance found within nature. This winter, I often questioned myself, do I have what it takes?, but maybe that is not the right question to be asking. Instead I need to ask myself if we are accomplishing the goals we set out to achieve in homesteading. Have I—have we as a family – learned the value of the natural world, the value of hard work, and the value of the food we produce and eat on this homesteading journey? The answer is undoubtedly, yes.

Acknowledgements

Our loving Creator, my Heavenly Father, who continues to amaze and inspire me, and who lets me rest in green meadows, gently leads me beside peaceful streams and who refreshes my soul.

My husband Nathan, aka, Mr. Green Thumb, for believing I could write this book, even on the days I didn't, and for always believing in me.

AJ, Aussie, Rae-Rae, and CJ, for giving me permission to share about their lives in this book, and for helping me in any number of ways so that I could finish writing.

My encouraging, supportive and patient publishers, Emily and Tom, for taking a chance on me, and helping me stretch and grow as a writer in the process. And Dr. Dave, for strengthening my writing with his gifted editing skills.

My parents, for reading many of these chapters with enthusiasm and encouragement as I wrote them, and for helping out on the homestead when we needed an extra hand.

My mother-in-law Maryann, for all of her gardening advice and plant starts, the original Ms. Green Thumb.

My brother-in-law Chris, of One Love Farm, who has taught me much about plants and gardening; and my sister-in-law Becky, for teaching me how to make soap, and always taking an interest in the book's progress.

My brother-in-law, Michael, who got me the writing gig that led to this book.

My friend Marlo, for regularly checking in with me to ask how my writing was going, and to give me a few good pep-talks when I needed them most.

All of our supportive friends and family who have taken an interest in this book, in our homestead, and more importantly, in our lives over the years.

Our gracious neighbors, who Scarlett would nip at whenever she got loose; and who mostly tolerated Mr. Peabody making a mess on their patio; who have never complained of the rooster's crows or any of the other sounds or smells wafting from our homestead.

Our loyal customers, you know who you are! We could not have gotten to this point in our homesteading journey without you buying our eggs and pork, and believing that ethical food raised sustainably is worth supporting.

Dawn of Peppinbrook Farm, whom I have learned much from in regards to poultry and animal husbandry. She also took in Rowan after losing his companions to coyotes. She has been my greatest homesteading enabler, and I cherish the birds and kittens we have received from her.

Rebecca of Yellow Point Farm, who sold us our stunning herdsire Rowan, and his brother, Gingko. And who gave me advice and support as I made some difficult decisions in regards to my goats.

Brian Giesbrecht, who we bought our first bacon seeds from, and who has generously helped us load and take our pigs each year to butcher, as we don't have a suitable trailer of our own.

Leona of Cedar Green Farm, who, although we have never met in person, was willing to coach me through managing Clover's prolapse over the phone.

Chelsey of Felling Family Farm, who we purchased our lovely meat goats from, and who we have purchased hatching eggs and chicks from over the years.

Judy and Mike Campbell of Campbell's Gold Honey Farm & Meadery, who hosted and taught the fabulous course I completed on keeping honey bees.

Heidi, my friend and beekeeping mentor, who patiently taught me about beekeeping as she worked on her hives, and has helped support me in my own beekeeping venture.

Brian Minter of Minter Country Garden Store, who so generously helped us design our orchard and select the trees most suitable for our climate.

Lepp Farm Market, who has done a fabulous job of processing our pigs.

Back to Basics Facebook group, a local, welcoming online community where I have learned about animals, gardening, fermenting, and any number of other hobbies I have attempted in relation to homesteading from others on their own homesteading journeys.

www.ingramcontent.com/pod-product-compliance
Lightning Source LLC
Chambersburg PA
CBHW042248240426
43672CB00020BA/2989